SPACE AND DINAMICS MASS-ENERGY

Licence CC0 ✓

AMAZON
2018, Luis Javier Artieda Carpio
ISBN 9781983041587
Independently published

All rights reserved
This publication can't be reproduced

DYNAMIC BALANCE

OR

UNIVERSAL EXPANSION

Author: LUIS JAVIER ARTIEDA CARPIO

TO

THE INNOCENT VICTIMS OF

IROSHIMA - NAGASAKI - CHERNOBIL

AND TO INNOCENT PEOPLE THAT SUFFER,

WITHOUT DEFENSE,

CATASTROPHIC EVENTS

CAUSED

BY THOSE WHO HAVE CAPTURED THE POWER

IN OUR WORLD

I BELIEVE IN HUMAN BEING

I believe that the human being is part of the creative process to which, long ago, he is in the dual position of observer and actor.

Paradoxically, its enormous work capacity and the scientific and technological advance have turned him into a potentially dangerous being for the Earth. The man's current power to alter the earth ecology has, as counterpart, his ineptitude or unwillingness to define, in advance, all the negative variables that unleash their industry, science and their belligerent and aggressive relationship against its peers in the absurd and constant search to impose their will.

One way to defend the Earth is to help man understand his planet and the fragile "DYNAMIC BALANCE" in which we live. Humanity must understand that it has discovered and developed means capable of altering that balance with harm for to those who play with that power with impunity. But also, unjustly, for all men, women, children and every living being that can't avoid tyranny of the powerful.

The balance in planetary systems requires:

Centrifugal Force (Fcf) = Centripetal Force (Fcp)

a) Dynamic Balance or Dynamic equilibrium

When talking about the nature we must accept that it is in permanent change; therefore, the concept of static equilibrium is insufficient. This leads us to the DYNAMIC BALANCE applicable to all systems in the universe, independent of their dimension.

Dynamic equilibrium is the state in which any alteration is compensated simultaneously by the re-adaptation of the system to a new state of equilibrium. Re-adaptation occurs in a time and a space; therefore time and space are interdependent and unavoidable factors.

In this work we will talk about the EME

ENTE MASS-ENERGY (EME)

[ENTE: From the Latin ENS, it is a philosophical concept that refers to what is, exists or can exist. An entity participates in being and has properties that, as a being, are its own. The concept transcends the matter, since an entity can be a galaxy, an atom, a table, a television, a lake or the square root of sixteen. (Wiki: Definition of entity (http://definicion.de/ente/))]

Luís Javier Artieda Carpio - Febrero - 2016

FOREWORD

EARTH IN DANGER

Genesis. - 2.4 to 9

וַיְכַ֤ל אֱלֹהִים֙ בַּיּ֣וֹם הַשְּׁבִיעִ֔י מְלַאכְתּ֖וֹ אֲשֶׁ֣ר עָשָׂ֑ה וַיִּשְׁבֹּת֙ בַּיּ֣וֹם הַשְּׁבִיעִ֔י מִכָּל־מְלַאכְתּ֖וֹ אֲשֶׁ֥ר עָשָֽׂה׃ On the seventh day God finished the work that He had been doing, and He ceased on the seventh day from all the work that He had done.

3

וַיְבָ֤רֶךְ אֱלֹהִים֙ אֶת־י֣וֹם הַשְּׁבִיעִ֔י וַיְקַדֵּ֖שׁ אֹת֑וֹ כִּ֣י ב֤וֹ שָׁבַת֙ מִכָּל־מְלַאכְתּ֔וֹ אֲשֶׁר־בָּרָ֥א אֱלֹהִ֖ים לַעֲשֽׂוֹת׃ (פ) And God blessed the seventh day and declared it holy, because on it God ceased from all the work of creation that He had done.

4

אֵ֣לֶּה תוֹלְד֧וֹת הַשָּׁמַ֛יִם וְהָאָ֖רֶץ בְּהִבָּֽרְאָ֑ם בְּי֗וֹם עֲשׂ֛וֹת יְהוָ֥ה אֱלֹהִ֖ים אֶ֥רֶץ וְשָׁמָֽיִם׃ Such is the story of heaven and earth when they were created. When the LORD God made earth and heaven—

5

וְכֹ֣ל ׀ שִׂ֣יחַ הַשָּׂדֶ֗ה טֶ֚רֶם יִֽהְיֶ֣ה בָאָ֔רֶץ וְכָל־עֵ֥שֶׂב הַשָּׂדֶ֖ה טֶ֣רֶם יִצְמָ֑ח כִּי֩ לֹ֨א הִמְטִ֜יר יְהוָ֤ה אֱלֹהִים֙ עַל־הָאָ֔רֶץ וְאָדָ֣ם אַ֔יִן לַֽעֲבֹ֖ד אֶת־הָֽאֲדָמָֽה׃ when no shrub of the field was yet on earth and no grasses of the field had yet sprouted, because the LORD God had not sent rain upon the earth and there was no man to till the soil,

6

וְאֵ֖ד יַֽעֲלֶ֣ה מִן־הָאָ֑רֶץ וְהִשְׁקָ֖ה אֶֽת־כָּל־פְּנֵֽי־הָֽאֲדָמָֽה׃ but a flow would well up from the ground and water the whole surface of the earth—

7

וַיִּיצֶר֩ יְהוָ֨ה אֱלֹהִ֜ים אֶת־הָֽאָדָ֗ם עָפָר֙ מִן־הָ֣אֲדָמָ֔ה וַיִּפַּ֥ח בְּאַפָּ֖יו נִשְׁמַ֣ת חַיִּ֑ים וַֽיְהִ֥י הָֽאָדָ֖ם לְנֶ֥פֶשׁ חַיָּֽה׃ the LORD God formed man from the dust of the earth. He blew into his nostrils the breath of life, and man became a living being.

8

וַיִּטַּ֞ע יְהוָ֧ה אֱלֹהִ֛ים גַּן־בְּעֵ֖דֶן מִקֶּ֑דֶם וַיָּ֣שֶׂם שָׁ֔ם אֶת־הָֽאָדָ֖ם אֲשֶׁ֥ר יָצָֽר׃ The LORD God planted a garden in Eden, in the east, and placed there the man whom He had formed.

9

וַיַּצְמַ֞ח יְהוָ֤ה אֱלֹהִים֙ מִן־הָ֣אֲדָמָ֔ה כָּל־עֵ֛ץ נֶחְמָ֥ד לְמַרְאֶ֖ה וְט֣וֹב לְמַאֲכָ֑ל וְעֵ֤ץ הַֽחַיִּים֙ בְּת֣וֹךְ הַגָּ֔ן וְעֵ֕ץ הַדַּ֖עַת ט֥וֹב וָרָֽע׃ And from the ground the LORD God caused to grow every tree that was pleasing to the sight and good for food, with the tree of life in the middle of the garden, and the tree of knowledge of good and bad.

The Bible places Adam and Eve in an Eden that they quickly lost.

The third millennium of our era finds Homo Sapiens-Sapiens in transition towards a profoundly different world in which, man has profoundly altered the vital functions of the biosphere.

In fairness we must recognize that in past, other massive changes altered the life of our planet. The continents have moved, appeared and disappeared seas and mountains, eras of heat and cold altered the global climate and extinguished successively; even living beings were and are factors of climate change because their metabolism helped the gradual enrichment of atmospheric oxygen and carbon consumption to transfer it to its own structure. Mass extinctions of animal and plant species have accompanied these changes, but life returned evolved.

However, in the last two centuries, an important qualitative difference has emerged. In its immeasurable need to usurp habitats in search of natural goods to transform them into raw material, without measure or limit, the human being has produced an artificial but irreversible period of extinction.

Today we know that environment pollutants materials are pushing the earth globe's atmospheric temperature to ever higher levels with irreversible damage to whole earth; although its only and debatable merit is to made more comfortable the first world's life, or enable the existence of companies whose profits are measured, only, in terms of annual economic profit.

On the other hand, huge and growing human masses raise unprecedented demands on farms, forests and other resources. The expansion of technological civilization has made possible the growth of the world population, but at the same time produced irreversible changes in the vegetation of the continents, and interacts with the atmosphere, continents, oceans, rivers, lakes, soils, subsoil. All this with the exclusive object of keep the humanity, in the same time that it produces the greatest living beings' extinction and their habitats which the planet never saw.

In parallel with its destructive potential, the human being has scientific and technological tools that have given him an unimaginable power in the past; Satellite-based sensors, sophisticated terrestrial facilities and macro computers that can analyze, store, assemble and evaluate millions of data simultaneously in fractions of second. Today, with these tools, human being is able to observe and understand the global mechanism of life, and could try to change what has been their unconscious play with the future.

"Can today's human being overcome his mythical father? Will he keep his Paradise?

Graphic # 1
DYNAMIC BALANCE OR UNIVERSAL EXPANSION
HUMAN BEING, OBSERVER AND ACTOR

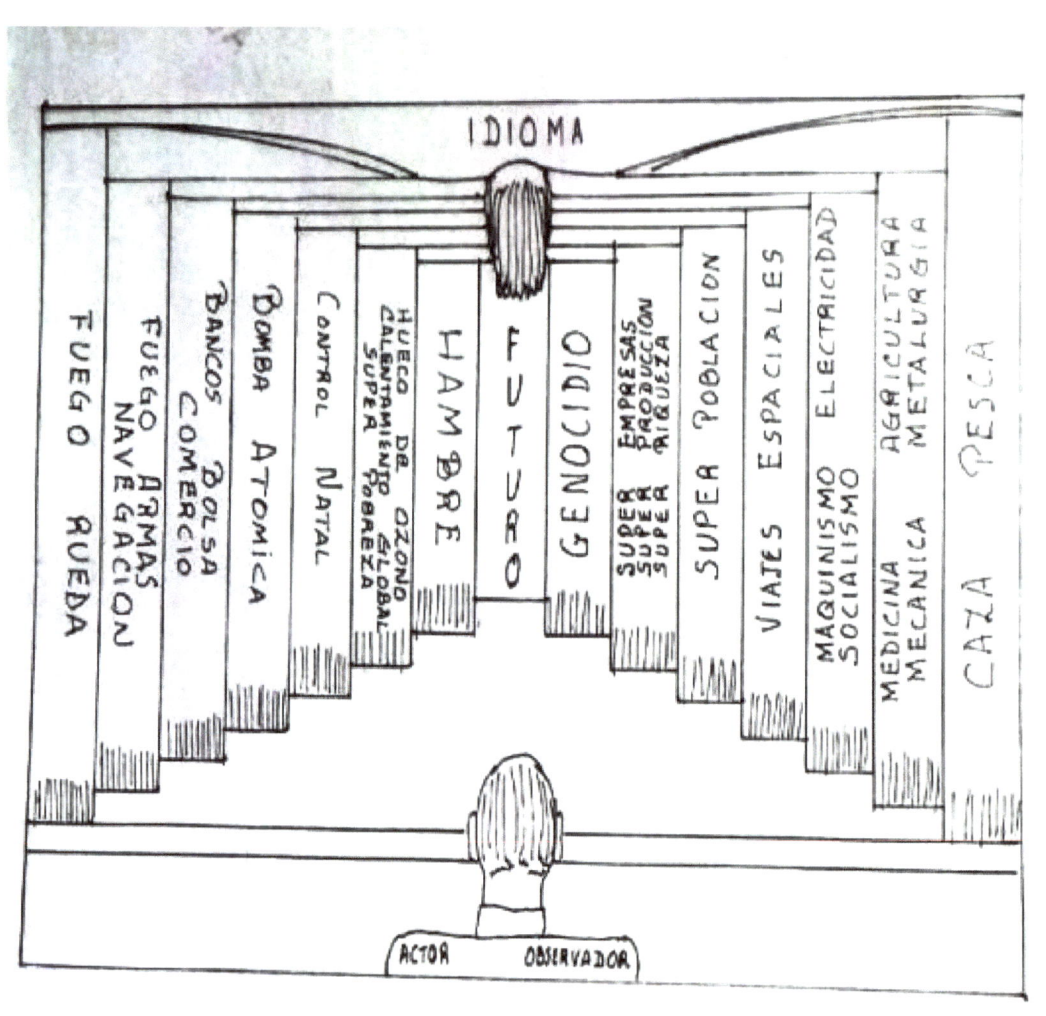

INDEX
BOOK ONE

TO THE INNOCENTS
I BELIEVE IN THE HUMAN BEING
FOREWORD - EARTH IN DANGER
BOOK ONE
DYNAMIC BALANCE OR UNIVERSAL EXPANSION
1) Equilibrium
2) Structure and Expansion of the EMEs
3) Universal Space (US) and EME (Energy-Mass Entities)
4) Mass Energy Systems
5) Formation and Expansion of a Planetary System
6) Solar System Expansion
7) Mass and / or Solar Energy Emission
8) Sun's Mass/Energy irradiated through its Life
9) Rotation of an EME
10) Solar System's Dynamics
11) Mass and Energy Emitted by Planets
12) The Solar System, does it expand?
   CONCLUSION

BOOK TWO
THE EARTH

EARTH, DOES IT EXPAND IN A DYNAMIC BALANCE?
1987 – EQUILIBRIUM OR EARTH DIAMETER EXPANSION
1) Hypothesis: Earth Expands Dynamic and Balanced
2) Ancestral Cultures Calendars
3) The Earth
4) Centrifugal Force Vs Centripetal Force
5) Radios of Spheres that grow
6) Chronological Attempt
7) The Continents and the Dynamic Balance
8) Magma Tectonic Inertial Runoff
9) Magnetic Belts or Magnetic Bands
10) The Continents and the Expansion
11) Terrestrial Structures of Similar Tectonic Development
12) America and Euro-Asian's cordilleras Systems
12) The Oceans
   CONCLUSION

BOOK THREE
EARTH THERMODYNAMIC EQUILIBRIUM

1) Discontinuities what are they?
2) Magnetic Field and energy leak
3) Tides and its effects
4) "El Niño" Phenomena
5) Human Action
6) Greenhouse Effect
7) Greenhouse Phenomenon Consequences
   EPILOGUE

GRAPHIC INDEX

BOOK ONE
Graphic # 1 Human Being Observer and Actor
Graphic # 2 Balance in Giving and Receiving
Graphic # 3 Solar System Data Table
Graphic # 4 Variation of a Planetary System
Graphic # 5 Sun Radiate Matter - Energy at Expense of Its Own Mass
Graphic # 6 Sun Expulsed Planets from Itself, Successively
Graphic # 7 Solar Planets, Data of Rotation and Orbital Period

BOOK TWO
Graphic #   1 Earth Expansion Process
Graphic #   2 From Paleozoic to our Days
Graphic #   3 Evolution of Arctic Ocean
Graphic #   4 Antarctic Continent Evolution
Graphic #   5 Oceanic and Terrestrial Mountain Chains Formation
Graphic #   6 Lung Fishes
Graphic #   7 Australia, Papua New Guinea, New Zealand
Graphic #   8 Mexican Gulf, Caribbean Sea and part of South America
Graphic #   9 Magma Tectonic Inertial Runoff
Graphic # 10 Magnetic Strips
Graphic # 11 Mountain Chains from Spain to China
Graphic # 12 Mountain Chains from Alaska to Patagonia
Graphic # 13 New Zealand and its Submarine Surrounding
Graphic # 14 Peninsula Indonesia, Malaysia, Indonesia, Philippines, Papua
Graphic # 15 Patagonia, Antarctic Peninsula, South Sandwich Islands
Graphic # !6 Terrestrial Structures of Similar Tectonic Development
Graphic # 17 Pacific Ocean and Beds, Northern, Center and South
Graphic # 18 Atlantic Ocean, Central, South and Beds
Graphic # 19A North Indic Ocean and its Beds
Graphic # 19B South Indian Ocean
Graphic # 20 Earth Crust, modified during Angular Speed Acceleration
Graphic # 21 Magma Arcs, product of the Earth Expansion
Graphic # 22 Insular Arcs product of Earth Expansion
Graphic # 23 Sea of China, Philippines, Japan and Ocean Floor
Graphic # 24 Coco's Plate
Graphic # 25 Arabic Sea
Graphic # 26 Antarctic Ocean and Bed
Graphic # 27 Arctic Ocean and Bed

BOOK THREE
Graphic # 1   The Earth, a Thermal Machine
Graphic # 2   Deforestation
Graphic # 3   Agriculture and Deforestation
Graphic # 4   Destructive Irrigation
Graphic # 5   Plastic garbage in the Oceans
Graphic # 6   Ocean Pollution

BOOK ONE

DYNAMIC BALANCE OR UNIVERSAL EXPANSION

1) EQUILIBRIUM

[EQUILIBRIUM: (Lat. Aequilibrium) The state of a body subjected to opposing forces, which compensate each other. - Balance state between concurrent forces on a given system, under these conditions the system does not undergo significant change.]

Planetary systems require:

Centrifugal Force (Fcf) = Centripetal Force (Fcp)

a) Dynamic Equilibrium

When speaking of nature we must accept that it is in permanent change, therefore, the concept of static equilibrium is insufficient. This brings us to the DYNAMIC BALANCE applicable to all the systems of the universe, regardless of their nature.

Dynamic equilibrium is the state in which any alteration is compensated simultaneously by the system's re-adaptation to a new state of equilibrium. The re-adaptation occurs in a time and a space; therefore time and space are interdependent and unavoidable factors

b) MASS-ENERGY Entity (EME)

Entity: From the Latin ENS, it is a philosophical concept referred to what 'is, exists or can exist'. It is an entity, part of being, with characteristics that are its own. An EME can be a galaxy, an atom or an atomic particle.

Entity: something that exists apart from other things, that exists independently (Examples: A galaxy, a planet, an atom, a neutrino, etc.)

We also know that every Mass-Energy Entity (EME) constantly gives part of its energy to the Universe; therefore, it is permissible to ask ourselves: what role does energy play in the great equation of the UNIVERSAL EQUILIBRIUM? (See Graphic # 2. - BALANCE IN GIVING AND RECEIVING).

In addition, by Albert Einstein, we know that mass and energy are interchangeable and, as stars have radiated an unimaginable amount of energy at the expense of their mass during their immeasurable existence of billions of years, we conclude that: the mass of all Stars, like our Sun, is smaller today than in the past; and it will continue diminishing with time.

GRAPHIC # 2
THE DYNAMIC BALANCE OR THE UNIVERSAL EXPANSION
BALANCE IN GIVING AND RECEIVING

The 'EQUILIBRUM' is possible IF 'GIVING' is 'EQUIVALENT' to 'RECEIVING'
Otherwise the 'DESEQUILIBRIUM' occurs
So one of 'SOURCES' go 'EXHAUTED'

The heliocentric conception (Sun center of the system) enunciated by Galileo and supported by Copernicus (contemporaries of Colon 1473-1503), corrected the old geocentric idea of Ptolemaic times (earth center of the system). Copernicus and Galileo discovered

Earth's and planets' movements of rotation and translation but, when the notable Italian sage made public his discoveries was condemned by heresy. Nevertheless, time gave him the reason and imposed their ideas with indisputable power. It is fair to acknowledge that Greek Aristarchus (n. 208 BC) on Samos, the Greek Aegean's island, was the first (known) to claim that Earth and the planets revolve about themselves and around the sun. (In his religious cult was also declared heretic).

Based on Tycho Brahe observations, Johannes Kepler defined the elliptical shape of planetary orbits and Isaac Newton proposed the "Universal Law of Gravity" and the "Motion Laws".

The data accumulation, properly coded, led to the fact that: Earth is the third planet in the solar system, she is a ENTITY MASS-ENERGY (EME) as all the planets, stars, asteroids, comets and others; is part of the chaos organized by and with billions constellations traveling in the Universe with unknown destiny, governed by forces not understood, yet. (See Graphic # 3. - TABLE OF DATA, SOLAR SYSTEM)

The integral vision of the phenomena and the advance of sciences has made possible the understanding of those phenomena and gradually erased the limits. Now we know; we are a tiny link in the universal phenomenological chain initiated in some remote corner of Cosmos where, in an immeasurable past, a hyper-energetic center was unleashed towards the inscrutable emptiness. The journey towards equilibrium is a consequence of the 'Energy Gradient' that sooner or later explains the simplest expressions of energetic transference. Heat flows to the cold, light to darkness, knowledge to ignorance, abundance to narrowness.

This first approximation tells us that; equilibrium is a form of transit towards zero energy level, towards absolute emptiness and immobility.

However, there is a gradation; phenomenological levels than non permit to recognizable the link between stages. The gradation refers to the unit of energy which, that in each case, is connected with the force that keeps, as a unity, the Energy-Mass Entity's (EME) parts. All that said is linked to the time during which phenomena occur.

c) Energy gradient

Every Mass-Energy's exchange phenomenon, between EMEs, is proportional to the existent potential energy difference. This energy potential difference modifies parameters such as: time in which change occurs, magnitude of change, phenomenological homogeneity and others. In other words, any energy transfer process requires gradient or energy differential that produces a form of transient dynamic imbalance that makes possible, limits, and governs change.

In other words; forced by the energy gradient, the Original Massive Center of Original Energy (EME-O) began its expansion towards the Universal Vacuum (VU), delivering enormous amount of galaxies and nebulae. At the same time; each of the Mass-

Energy Entities (EMEs), born in the original explosion, continued to expand attracted to the (VU) and generating myriads of stars in their own space and time.

The stars, in turn, deliver mass and energy in multitude of forms; Light, heat and matter. Colder sidereal bodies deliver it through thermal, electrical, magnetic, chemical, physical, biological, and other phenomena.

d) Universal Space

The UNIVERSAL SPACE (EU) is the stage for cosmic dimension events. Is a conjunction of unlimited void with a gigantic number of Mass-Energy Entities (EME), this conjunction interacts constantly. That is; the Energy- Mass of the EMEs will be absorbed into the Universal Vacuum (VU) until the disappearance of potential Gradient (under the assumption that, in an unpredictable future, this could be possible so). A limited space would have been saturated in a short time, but it was not. In other words, space is infinite and immeasurable.

GRAPHIC # 3
THE DYNAMIC BALANCE OR UNIVERSAL EXPANSION
SOLAR SYSTEM DATA TABLE

	Merc	Venus	Earth	Mars	Júpiter	Saturn	Uranus	Neptune
Diámetro Km	4870	12104	12756	6790	142800	119300	47100	48400
Dist.to Sun Mill Km	57.9	108.2	149.6	227.9	778.3	1427.0	2869.6	4496.7
Period. Sid Days Earth	87.9	224.7	365.26	686.9	11.8 años	29.4 años	84. años	165 años
Period Rot Days Earth	58.9	243.0	23.9 hr.	24.6 hr.	9.8 hr.	10.2 hr.	17.0 hr.	16.11 hr
Long Orbit Mill Km	363.8	679.8	940.0	1432.0	4890.3	8966.2	18030.4	28252.3
Orbit speed Km/seg	48.0	35.4	30.2	24.3	18.3	1208	8.5	7.1
Planetary Declination	28°	3°	23.27°	23.59°	3°05	26°44	82°05	28°48
Ecuad run mill. Km	365.1	688.4	954.6	1446.6	6823.8	11917.1	22570.4	37400.6
Periferic speed Km/seg	0.003	0.0018	0.4638	0.2408	12716	10207	3806	2673

On the other hand it is said that, in the Universe, the amount of matter-energy remains constant even when there are exchanges between Entities Mass-Energy (EME). In other words, the interaction between mass and energy produces obvious changes without thereby altering the great matter-energy quantum. We can repeat here, "Neither matter nor energy can be created or destroyed, they can only be transformed".

e) Final Balance

In his eagerness for increase their knowledge, the human being has lifted one after another the curtains of the unknown and has reached limits not dreamed; Nevertheless it is still looking out into the future through windows in whose limitless horizon there are galaxies, constellations, black holes and other giant phenomena; While at the other end nuclei, electrons, protons, diverse rays, photons, neutrinos and now Higgs' Bosons mark a fatuous limit that augurs unknown fields beyond the always astonished human imagination.

2. - THE EMEs, ITS STRUCTURE AND EXPANSION

Every EME is subject to the universal energy gradient, has structure and expands

a) **Structure**: System Focus, Satellites, System's atmosphere

- System's Gravity Center: the focus of the system is also an expanding EME; it occupies the system's geometric center, and contains a very large percentage of the system's Mass-Energy

- Satellites: they are EMEs that rotate in diverse orbits around the focus. The mass-energy of the satellites is small in relation to the focus. They are held in orbit by the force of attraction that is in direct relation to the masses of the focus and the satellite and inverse to the distance that separates them.

- **Atmosphere of the system**: It is born from the Mass-Energy of the systemic Focus, flowing away from the Focus at great speed. It surrounds the systemic focus, decreases in density as the distance to the focus increases. It is formed by sub-atomic particles. It expands constantly in response to Universal Gravitation (GU). It sweeps similar matter from atmospheres of the satellites

b) **Expansion**: Volumetric growth process of all EME, responding to Universal Gravitation

- All EME will expand as long as there are energy gradient between the EME and its cosmic atmosphere. Each EME delivers Mass-Energy to its cosmic atmosphere, extracting them from itself.

- Universal gravitation acts on EMEs and produces a gradual decrease in their Mass-Energy. The EME's internal gravity, which acts in opposition to the Universal Gravitation, decreases constantly.

- The expansion affects the EME throughout its structure. That is, each EME matter unit moves away from the center, but also from its own center.

- Every EME will decrease in density as it expands integrally.

3. - UNIVERSAL SPACE (EU) AND EMEs (Energy-Mass Entities)

Every EME occupies a space in the Universe. In Universal Space or Universal Vacuum (VU) all EMEs are in permanent expansion, as well as each and every one of its parts, regardless of its dimension.

a) Forces that govern the Dynamic Equilibrium on EME systems

In a cosmic system of any dimension, composed of System Mass' Focus and one or several Captive Masses (satellites) that revolve around in certain energetic levels (orbits), the equilibrium depends on the forces applied to the system. The dynamic equilibrium between Focus-Mass and Captive-Mass depends, from moment to moment, on the relationship between the Centripetal Force (System Internal Gravity CpF) and the Centrifugal Force (Cosmic Gravitation CfF) affecting them.

We know that, over time, all EMEs lose 'Mass-Energy'. Consequently its gravitational power decreases and the system expand evenly. Likewise in that system, the rotation and orbital velocity of EME satellite vary according to the dictates of the laws of Physics discovered and enunciated by Kepler and Newton.

- **Centrifugal Force Factors (CfF)**

As established by the Law of Universal Gravity discovered by Isaac Newton; for every satellite that rotates around a cosmic system (Mass-Energy), the Centrifugal Force (CfF) is proportional to the product of satellite's orbital speed multiplied by the dimension of its mass.

If the centrifugal force needed to keep an object moving in an orbit increases, it is because one or both of the following events occur:
The mass of the object increases
The speed of the object increases
Consequently, the balance, understood as the maintenance of the distance between the satellite and its Focus-Mass, depends on the degree of variability of the mass of the satellite and its orbital speed.

- **Centripetal Force Factors (CpF)**

For the same satellite, the Centripetal Force or attraction force (FCp) is proportional to the dimensional product of Focus mass by the mass of the satellite, and inverse to the distance that separates them. Consequently the equilibrium (understood as maintenance of the distance that separates the satellite from its Focus) depends on the variability of the masses and the distance that separates them.

The jump from one level of equilibrium to another (orbit) demands increase or decrease mass-energy.

In Space all EME radiate energy and / or matter continuously; consequently its expansion is continuous and unavoidable.

Any decrease in mass leads to a form of imbalance that is compensated by changes in the size of the orbits and / or orbital velocities of the satellites of the system. These modifications restore the dynamic balance to the system and return equality to the equation: Centrifugal force = Centripetal force.

In other words the system expands in a balanced way with and as a result of the loss of mass-energy and in response to Universal Gravitation.

The SPACE in which these events are realized is infinite, immeasurable and unitary.

From this it follows that: there is no natural energy source, sufficient and capable for to restructuring the Energy-Mass to any EME, or to restore its original condition of equilibrium and, much less, to contract to its original condition the universe. Consequently, and independently of its dimension, all EME sidereal systems expand permanently without possibility of contraction in the infinite space.

b) Changes in an EME

With time and the uninterrupted process of expansion, the energy gradient becomes smaller. This condition will lead to the death of the EME, under the assumption that would be possible to reach the "zero energy" level; at the end of infinite expansion through an infinite time!

In consequence:

- The law of universal attraction is applicable to all matter, whatever its state and dimension.
- Every spatial system expands constantly but maintains dynamic balance inside itself and with its environment.
- In the Universe, any expansion process is related to the energy gradient that originates the change, and also the forces that intervene and accelerate or delay the process.
- In the Universe, any expansion process is carried out in a time related to the dynamic imbalance that has generated the phenomenon.
- In the universe, any alien acceleration to dynamic equilibrium of a process would alter the dynamics of equilibrium and produce unpredictable results.
- The each unit EME's mass, regardless of its physical volume, becomes gradually smaller in response to the energy gradient that exists between its massive center and its outer crust.

- The constant mass-energy decrease of the EMEs would lead to the mass-energy = zero condition; If and only if it were possible to reach that utopian condition.

4) MASS-ENERGY SYSTEMS

EMEs of any size (galaxies, stars, planets, atoms, subatomic particles, etc.) are space entities that concentrate mass and energy in various forms and deliver them through constant expansion processes in response to the dynamic-energy gradient, Originated by Universal Gravitation. (See Graphic # 4 VARIATION OF A PLANETARY SYSTEM)

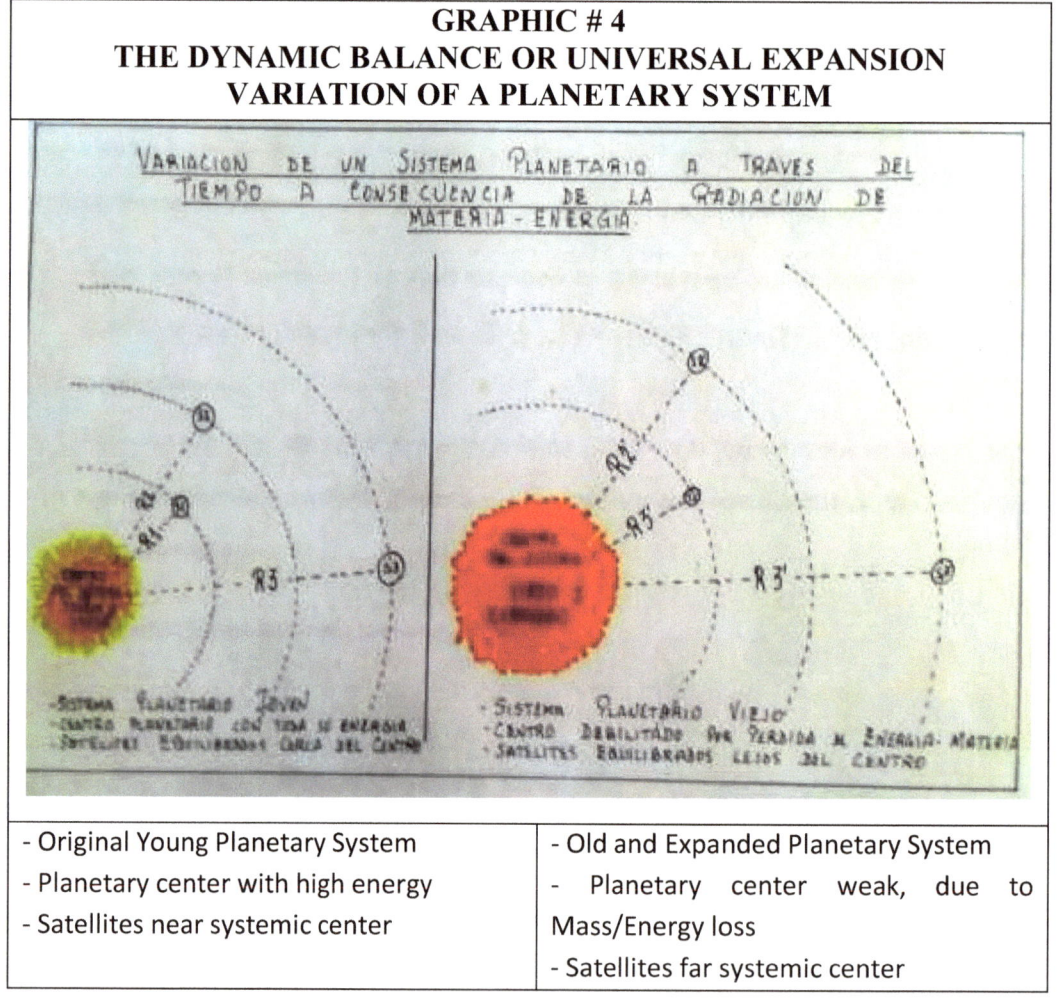

- Original Young Planetary System - Planetary center with high energy - Satellites near systemic center	- Old and Expanded Planetary System - Planetary center weak, due to Mass/Energy loss - Satellites far systemic center

a) Balance in a Sidereal System

Any sidereal system, with Mass-Focus (MF) that possesses the largest percentage of the mass of the system accompanied by smaller EMEs, is constantly subjected to Universal Gravitation and, in turn, generates enough force of internal attraction. The equivalence between Universal Gravitation and internal attraction will keep the system in a dynamic equilibrium

This is verified by studying the relationship between a planet mass and its Focus Mass (MF) around which it rotates (Example: Earth around the Sun). The relationship between the Focal mass with each satellite masses is subject to Kepler's Laws and Universal Gravitation discovered and enunciated by Isaac Newton. From all this we can deduct:

- Every planet (such as Earth) remains in orbit around its MF (Sun) because centripetal force (FA) and centrifugal force (FC) are constantly equivalent.
- According to the Movement Laws (Newton), sidereal systems possess dynamic equilibrium's mechanisms that constantly compensate the variations that occur.

These statements can be expressed as:

1.- $Fa = Fc$

Where: Fa Centripetal Force (attraction)
Fc Centrifugal Force

The formulas for Fa and Fc are

2.- $Fa = GMm / r^2$
3.- $Fc = 1/2\, mv^2$

From the formulas 1, 2 and 3 it follows that the mathematical expression of equilibrium in sidereal systems is:

4. $GMm / r^2 = 1/2\, mv^2$

Where:
G = Universal gravitational constant
M = Systemic Focus Mass (MF)
m = mass of the planet
r = Distance between masses M - m
v = Translational speed of the planet

From 4 it follows that

5.- $$\frac{v^2 r^2}{2GM} = \frac{m}{m} = 1 = EQUILIBRIUM$$

In other words people born before World War II learned and accepted, from Kepler and Newton, that the EQUILIBRIUM CONDITION of planetary systems was unalterable. However, Einstein's relativity theory proved that the parameters of equation (5) are subject to a constant variation. According to the theory of relativity, today we can affirm that light, stellar wind (like the solar wind) and the heat of stars come from a gigantic transformation of stars mass (M) to energy.

From this 'realization' we deduce that equation (5) has at least the 'independent variable' "M" (The system star's mass) whose value decreases constantly. The decrease of "M" constantly breaks the equilibrium of the system and equation (5) changes to:

6.- $$\frac{m}{m}<\frac{r^2 v^2}{2G(M-\partial M)}>1 = UNBALANCE$$

In which (∂M) is the mass transformed into energy and released into space by the massive center. As a result of the imbalance produced by that transformation, the system could be destroyed by flight of the planets with increasing acceleration. However the system is not destroyed because other variables change when "M" decreases.

b) Dynamic Balance Restitution Process

In the complex process of 'Dynamic Equilibrium', the 'Dependent variables' are: Distance between masses (r), planet's orbital speed (v). The process would be as follows.

- Initially, the Mass-Focus mass' decrease that goes, from (M) to (M-∂M) affects the force of attraction (Fa), which decreases to (Fa -∂Fa). Consequently, the distance (r) of the satellite tends to increase to $(r + \partial r)^2$.

- The system compensates for this tendency by the physical principle that governs the Kinetic Energy (Ec) of bodies in motion that says: "The (Ec) of a body in motion is constant before and after any change of velocity or direction". That is to say, to compensate for the variations of (M) and to comply with the physical principle described, the orbital velocity (v) becomes (v-∂v). This decrease is proportional to the decrease of the mass of the central body (∂M) and to the increase of the distance (r + ∂r) which rebalances the system constantly.

7.- $$\frac{(v-\partial v)^2(r+\partial r)^2}{2G(M-\partial M)} = 1 = BALANCE$$

It is important to note that the new equilibrium is based on different dimensions of (M), (r) and (v). That the percentage variation of (M) is very small, the increase in distance between the massive center and the satellite is very small, and the decrease in the planet's orbital speed is comparatively larger. See 'Orbital Speed of Planets' in Graphic # 3 (Solar System Data Table).

c) Time and Balance

As a consequence of the decrease in (M) we find that the equilibrium of the system persists because dependent variables (v) and (r) change in value. However that process happens through certain time.

- If the mass decrease (M) were 'significantly large' and the process occurred in 'significantly small time', galaxies, stars, planets, satellites and all sidereal entities would move away from their massive centers at explosive speeds, but this is not so . Consequently (∂M) is so small (from the human temporal perspective) that it produces no perceptible change (in human time). Therefore, we can conclude

- Space bodies constantly lose mass as a result of Universal Gravitation.
- The loss of mass, both in the mass center (M) of the system and in its satellites (m), produces changes in dependent variables such as orbital speed (v), rotation speed (ω), distance to mass center (r) and others.
- The loss of mass affects the density of the (EME) massive space body that will tend to expand its volume.
- Regardless of its size, the balance of any spatial system is dynamic and varies permanently over time.

5) FORMATION AND EXPANSION OF A PLANETARY SYSTEM

There are not enough data on the formation of EME sidereal systems, so we must imagine hypotheses that explain it. (Graphic # 3 SOLAR'S SYSTEM DATA TABLE)

Hypothesis 1. - A planetary system is born from a nebula, pastoral masses and accretion matter.
- I- A nebula with a high level of energy, surrounded by shepherding masses (planet Embryo) those revolve around the energetic center of the nebula. Pastoral masses attract gaseous matter that surrounds the energy center of the nebula, in that way the Proto-Planets are consolidated by capturing gaseous matter by accretion. During this stage, the shepherds' rotation time is indeterminate
- II- The shepherds' speed variation during the time when they are transforming into planet is not related to the increase of their mass per accretion.
- III- During system formation time, the system energetic center retains its mass

Analyzing hypothesis I, applied to the solar system, can deduct that the planets' current rates of orbit speed and rotation speed are incompatible with this hypothesis, so that:

- The accretion radically changed the rotational and translational speeds,

Or

- The EME system was formed in different ways.

Hypothesis 2.- Planetary systems are formed when massive bodies, passing near a star, are captured by the star's gravity.
- I- The gravitational attraction of the star must be properly strong.
- II- The trajectory of the massive body (its center of gravity) must pass within the attraction distance of the star to allow its capture and maintenance in orbit around the star.
- III- The massive body's trajectory should roughly coincide with the ecliptic of the star.

IV- The discovery of planets in other galaxies, which occurred in recent years, forces us to think that: to feed all galaxies and their myriad stars, an infinite source of Proto-Planets would be needed, traveling through space in search of their own star.

It is clear that the foregoing possibilities are very small, so that:

- The trajectory characteristics of approaching, from the source of Proto-Planets and the star, would have to be absolutely strict to allow the capture of the Proto-Planet in orbit.

or

- EME systems are formed in different way.

Hypothesis 3. - A planetary system is formed from an expanding Energy Center. The system's satellites are born by expulsion of mass extracted from the Energy Center.

We will analyze this hypothesis using the solar system as an example. (See Graphic # 4 VARIATION OF A PLANETARY SYSTEM).

6) SOLAR SYSTEM EXPANSION

From the science book "Earth in the Solar System"; used by Middle School students in United States, and from its article "The Sun Retinue" I have taken the following paragraph.

"The structure of solar system is typical of stars that formed in isolation. As a young and hot star she launched material to the outside, the inner planets (Mercury, Venus, Earth and Mars) were released as small rocky bodies. On the contrary, the outer planets (Jupiter, Saturn, Uranus and Neptune) retained their gases becoming gaseous giants"

What is absolutely evident is that:

a) **Sun radiates Energy.**

> IS THE SUN LOSING MASS?
> In Sun by Brian Koberlein
> The Sun loses mass in two major ways. The first is through solar wind
> The second way the Sun loses mass is through nuclear fusion. The Sun fuses hydrogen into helium in its core……….. we find the Sun loses about 4 million tons of mass each second due to fusion.

An important part of the scientific community gives the sun 10,000 million years life, during which time the Sun has radiated energy.

Consequently, we can say that total Sun energy radiation has been gigantic. Today it is estimated that the sun's radiation reaches 10^{33} ergs per second, but it is also known that solar activity includes phenomena such as Solar flares capable of emitting a thousand times more energy per second than the entire sun during a time that usually reaches the 30 minutes, from a very small area.

It is inescapable to say that during the (hypothetical) 10,000 million years of life, the energy gradient between Sun and its surroundings has gradually diminished. This leads us to conclude that the radiation was comparatively greater in past, such as is evident from paragraph "The Suns retinue".

b) Solar Wind

In the early 1960s, it was confirmed that the sun's crown expanded continuously creating a "wind" of particles that turned out to be fundamentally:

... ionized hydrogen (i.e., protons and electrons); The total flow was radio spherical towards the outside, very variable in speed, but generally between 350 and 800 kilometers per second; Had an average density of 5.4 ions per cm3 and an ionic temperature of 160,000 Kelvin. (Fragment taken from the Art: "Interplanetary Particles and Fields". J.A. VAN ALLEN)

This discovery confirmed that from the our star's birth, solar radiation included matter in form of ionized atoms added to the thermal energy, light energy and all the range of rays that the sun has sent out into Space.

Further studies based on achievements through artificial satellites and other means such as coronal ejection spectroscopic observations of EMEs (Mass-Energy Entities), of stars other than the Sun, have confirmed what was discovered through 1960-1970. However, the data obtained and above all, the interpretations of those data have left more questions than certainties, as is clear from articles published in scientific journals. Below are some paragraphs from those articles.

In 1958, the Explorer I satellite discovered the Van Allen belts regions of radiation particles trapped by the Earth's magnetic field. In January 1959, the Soviet satellite Luna 1 first directly observed the solar wind and measured its strength. Feldstein, Y. I. (1986). "A Quarter Century with the Auroral Oval, EOS". Trans. Am. Geophys. Union. (40): 761. doi:10.1029/eo067i040p00761-02. Paul Dickson, Sputnik: The Launch of the Space Race. (Toronto: MacFarlane Walter & Ross, 2001)
Stella coronal mass ejections
From Wikipedia, the free encyclopedia
There have been a small number of CMEs observed on other stars, all of which as of 2016 have been found on M dwarfs. These have been detected by

> *spectroscopy........ The observed projected velocities of CMEs range from ≈84 to 5,800 km/s (52 to 3,600 mi/s). Compared to activity on the Sun, CME activity on other stars seems to be far less common.*
> *("Hunting for Stellar Coronal Mass Ejections" - Korhonen, Heidi; Vida, Krisztian; Leitzinger, Martin. "Dynamics of flares on late-type dMe stars: I. Flare mass ejections and stellar evolution"- E. R.; Foing, B. H.; Rodonò*
> *"A search for flares and mass ejections on young late-type stars in the open cluster Blanco-1" - Monthly Notices of the Royal Astronomical Society.*

From the last paragraph it would be possible to deduce that the speed of EMEs (CMEs in the article) reaches speeds higher than the solar escape velocity. This is a clear contradiction with other opinions.

> The elemental composition of the solar wind in the solar system is identical to that of the solar corona: 73% hydrogen and 25% helium, with some traces of impurities. The particles are completely ionized, forming very dense plasma. In the vicinity of the Earth, the solar wind speed varies between 200 and 889 km / s, with an average of about 450 km / s. The Sun loses approximately 800 kg of matter per second in the form of a solar wind.

From the diversity of data extracted from many articles it follows that scientific organizations are still far from discovering the real data. Likewise, discordant explanations given to phenomena such as 'ejection velocity', 'escape velocity', 'ejected mass average' and others leave room for hypotheses than those raised in the reproduced articles.

c) Could the Sun radiate planets or large masses?

Once again we take a paragraph from the article "The Sun Retinue"

"The Sun, like a hot young star launched material into space, the inner planets (Mercury, Venus, Earth and Mars) were released as small rocky bodies, unlike the outer planets (Jupiter, Saturn, Uranus and Neptune), which conserved their combustible gases becoming huge "gaseous giants"

> *Coronal mass ejection*
> Additionally we find in 'Wikipedia, the free encyclopedia'
>
> **Coronal mass ejection (CME)** is a significant release of plasma and magnetic field from the solar corona. They often follow solar flares and are normally present during a solar prominence eruption. The plasma is released into the solar wind, and can be observed in coronagraph imagery.
>
> Coronal mass ejections reach velocities from 20 to 3,200 km/s (12 to 1,988 mi/s) with an average speed of 489 km/s (304 mi/s), based on SOHO/LASCO measurements between 1996 and 2003

> IS THE SUN LOSING MASS?
> In Sun by Brian Koberlein
> The Sun loses mass in two major ways. The first is through solar wind. The surface of the Sun is hot enough that electrons and protons boil off its surface and stream away from the Sun, generating a "wind" of ionized particles. When those particles strike Earth's upper atmosphere they can produce aurora. The solar wind varies a bit in intensity, but from satellite observations we know that the Sun loses about 1.5 million tons of material each second due to solar wind. The second way the Sun loses mass is through nuclear fusion. The Sun fuses hydrogen into helium in its core………. We find the Sun loses about 4 million tons of mass each second due to fusion.

In the excerpt from the book "The Sun Retinue" we reproduced earlier, the author states that the planets were expelled by a young Sun in a simultaneous event. However, the natural diminution of the solar mass, which occurred through its long life; the distance of the giant planets (Jupiter, Saturn, Uranus and Neptune) as opposed to the small ones (Mars, Earth, Venus and Mercury); additionally to the planets' speeds diversity in their orbital and rotational movement; it all demonstrates that each planet was expulsed in an unique event and at different time.

7) MASS AND / OR SOLAR ENERGY EMISSION
THE SUN EXPULSES ENERGY AND MATTER PERMANENTLY:

Matter - Solar wind composed by Ions
Speed - 350 to 800 Km / Sec.
Density - 5.4 Ions / cm3 (when passing near the Earth)
Radiation - Radio waves, Infrared light, Visible light, Ultraviolet light, X-ray, other

If the Sun has expelled and expels ionic matter, could it have expelled Planets?

a) THE SUN HAS EXTRACTED ALL THE PLANETS FROM ITS OWN MASS AND PUTED THEM IN ORBIT ONE BY ONE.
The planets were born from the solar mass as a result of alterations of mass/energy that happen in Sun's interior, generated by asymmetries of the universal attraction or big thermodynamic differences in the Sun's mass.

b) EMPIRICAL ASSUMPTIONS
- The original Sun's mass is the sum of:
  - Current Sun's mass
  - 100 times the current mass of the planets

- Radiation of 10^{33} ergs per second in 10^{10} years
- Solar wind in 10 billion years
- Mass lost by each planet during its own life.

- The planets were put into their initial orbit by consuming an amount of solar mass, transformed into energy, equivalent to 100 times the original mass of the ejected planet.
- The planets were putted in orbited successively.
- Each planetary birth reduced the mass of the sun and, consequently, also reduced the attraction that it exerted on the planets in magnitude equivalent to the loss of mass.
- With each successive planetary birth, all planets increase their distance from the Sun, reduce its orbital velocity and increase the rotation speed in proportion to the decrement in mass suffered by the Sun.
- The solar system expansion follows the dictates of universal attraction laws discovered, interpreted and enunciated by Keppler and Newton and amplified by Einstein through his theory of relativity.

The first to be born were the dwarfs' planets like Pluto. Considering their orbit's characteristics, it is possible to suppose that, Pluto was born from the Sun in a time near or simultaneous to Neptune's born and influenced by its gravitational attraction. This fact sealed forever the Pluto's life. Next in born was Uranus. *(See Graphic # 5 Suns irradiates matter and energy at the expense of its own mass)*

Subsequently, a mature Sun in all its power, gave birth to the giants Saturn and Jupiter. Asteroids Belt between Jupiter and Mars would be a sign that the energy gradient between Sun and its surroundings went through a minimum energetic level stage.

The birth of Mars suggests a certain energy recovery that is ratified with two medium-sized planets and similar characteristics, Earth and Venus. Mercury is the latest medium activity of the Sun.

Taking as a starting point the Sun's present mass and evaluating the various 'emission of matter and energy' forms such as: birth of planets, all kind of energy radiation, solar wind, glows, and others; it is possible to empirically infer the mass loss suffered by our star in the life time that science assigns to the Sun.

In order to calculate the Sun's mass at its birth; it was taken as solar life time what science assigns to our star (10,000 million years); and it has been considered (empirically) that the amount of energy needed to put a planet into orbit is a hundred times the current mass of that planet. With these elements, it has been concluded that loss Sun's mass during its 10,000 million years, has been the most important factor in the dynamic balance of the solar system.

NOTE: It has been considered congruent to apply a multiplier factor for put planets orbiting since it is obvious that the "launching" effort occurs in conditions of very high inefficiency compared to the artificial satellite launches made by the human being. The factor used is arbitrary.

Another important issue to consider is; the Sun, was and still is in energetic capacity to expel, from itself, and put orbiting planets in the system? In our logical exercise this question has been one of the most difficult points to overcome. According to today scientific opinion; THE MINIMUM SPEED TO ESCAPE THE SUN's ATTRACTION OF THE SUN AND KEEPED TRAPPED IN ORBIT SOLAR, IS 618 Km / sec.

This speed is considered an insurmountable barrier and equivalent to the concept of "sound barrier" from the 1950s to the 1960s. In scientific texts it has been commented: "And the escape speed of the Sun? Applying the radius and the solar mass comes out of about 620 km / sec. With this speed (it is said) not even hydrogen or helium can escape ".

Under the influence of artificial satellite launches technology, we have forgotten an absolute fact; since the middle of the last century we know that the solar wind "blows" through the whole system and it moves uncontrollably towards the outer space with a speed bigger than620 km / sec. The scientific world has accepted the incontrovertible existence of the solar wind; therefore, it must also accept that the Sun has the capacity to expel of itself matter and energy at a speed superior to its escape speed. (The Sun is not a Black Hollow).

I claim the fact that the Solar Wind reaches speeds in excess of 3200 km / s. In the paragraph that I reproduce below, the scientific world today accepts that "Solar Coronal Mass" is ejected at that speed.

Coronal mass ejections reach velocities from 20 to 3,200 km/s (12 to 1,988 mi/s) with an average speed of 489 km/s (304 mi/s), based (Art. Coronal mass ejection From Wikipedia, the free encyclopedia

Nor should we ignore that E.N. Parker has stated that: "A glare can last 30 minutes and emit a thousand times more energy than the whole of the Sun from an area of only one ten thousandth of its total surface". (See Graphic # 6 In the Sun photo we can see three solar prominences of gigantic dimension; nothing prevents that these prominences reach greater energy and much less if we think in a younger Sun).

For all the above, it is possible to affirm that the planets came out of the Sun's body in successive and gigantic mass/energy births.

8) SUN's MASS/ENERGY IRRADIATED TRHOUGH ITS LIFE

As we have said the Sun has radiated mass/energy as: SOLAR WIND, BIRTH OF PLANETS and ENERGY WAVES, from the beginning through now, and will be same in future.

GRAPHIC # 5
DYNAMIC BALANCE OR UNIVERSAL EXPANSION
SUN RADIATE MATTER - ENERGY AT EXPENCE OF ITS OWN MASS

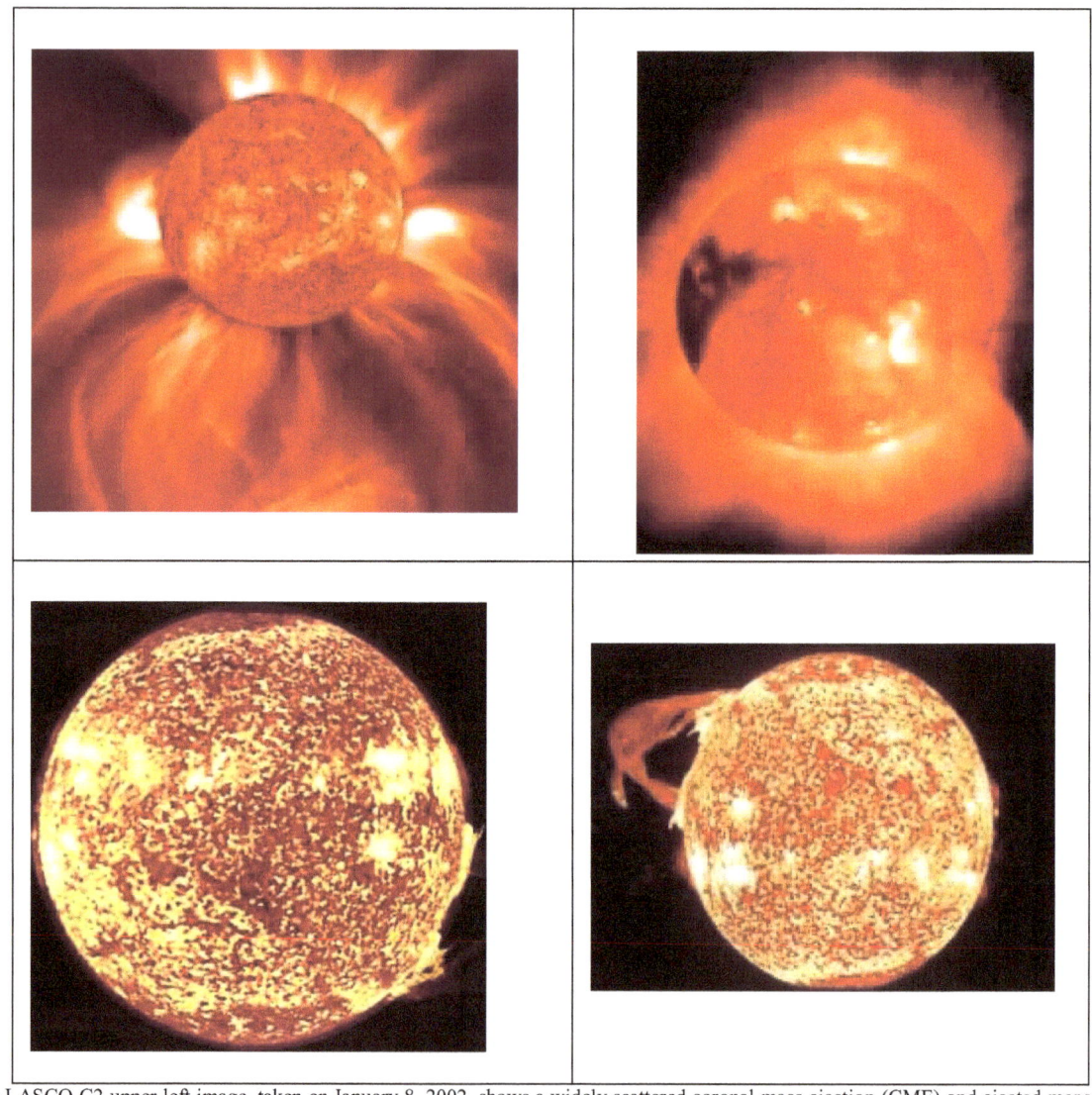

The LASCO C2 upper left image, taken on January 8, 2002, shows a widely scattered coronal mass ejection (CME) and ejected more than one billion tons of matter out into space at millions of kilometers per hour. These images were taken from GOOGLE EARTH. (http://meioambiente.culturamix.com/natureza/o-que-tem-dentro-do-sol.) (http://www.portalciencia.net/enigmamund.html.)

a) Solar Wind
J.A. VAN ALLEN.

"The medium through which the earth moves in orbit around the sun contains about 10 particles of matter per cm3: 5 positive ions (mainly protons) and 5 electrons. In comparison there are about 27x1018 particles in the same volume of terrestrial atmosphere at sea level. The basic characteristics of this wind are: variable in speed but in general between 350 and 800 km / sec, it comes out of the sun in a radial way and has an ionic temperature of 160º Kelvin when passing through the earth at present ".

> **Solar Wind's Composition**
> The solar wind's elemental composition in the system is identical to that of the solar corona: 73% of hydrogen and 25% of helium, with some traces of impurities. The particles are completely ionized, forming very dense plasma. In the vicinity of the Earth, the solar wind speed varies between 200 and 889 km / sec, being the average of about 450 km / sec.

Solar Mass Irradiated As Solar Wind

Calculating the spherical crown volume (centered in the Sun) generated in one second, by a balloon that grows at a speed of 800 km / s when passing through the earth, we will find a volume of 224.99×10^{33} cm3 full of wind solar at the rate of 10 particles per cm3 or approximately 2.25×10^{36} particles lost by the Sun as a result of the solar wind, in every second for 10,000 million years, which scientific world assigns to the Sun.

If it were accepted that the loss has been constant (not a real thing since the sun has lost the largest amount of particles), we must accept that in the time that has passed a "constant sun" would have lost at least 2.25×10^{36} ions/second by 3.1536×10^{17} seconds. = 7.0956×10^{53} ions/hydrogen. It is approximately $1,188 \times 10^{27}$ kg of mass by effect of the solar wind, at the actual rate.

This mass amount is almost as great as the sum of the present masses of all the planets as we see later.

b)   Birth of a Planet

Solar Mass Lost By Planets Birth.
The current mass of the Solar System is the sum of all EME masses in the system (see LAMINA 6. - The Sun, has expulsed from its own mass the planets, successively?

Current mass of the SUN = 1.9671052×10^{30} Kg

If, as a premise, it is accepted that put in orbit a Sun's EME, it demands to transform into energy a mass equivalent to one hundred (100) times the mass of the EME; the previous data should indicate that the energy consumed by the Sun to give birth to all the planets, it was at the expense of a significant percentage of the initial Sun's mass.

The loss of solar mass, because of planets, could have been: 2.64101518 x 10^{29} Kg.

TABLE of Current mass of Planets

MERCURY	0.3344 x 10^{24}
VENUS	4.8194 x 10^{24}
EARTH	5.98 x 10^{24}
MARS	0.62947 x 10^{24}
JUPITER	1877.995 x 10^{24}
SATURN	562.3954 x 10^{24}
URANUS	85.766 x 10^{24}
NEPTUNE	103.070 x 10^{24}
PLUTO	0.018 x 10^{24}
Total mass of all planets, today.	**2.6410077 x 10^{27}**

The percentage of solar mass consumed to give birth to the planets must have been:
11.836652% of the initial solar mass.

c) RADIATION
Solar Mass Lost By Radiation.

In his interesting article 'THE SUN', E. N. Parker states that

"…the enormous energy radiation of the sun (10^{33} ergs per second) is not due only to the combustion of a gas"; and after.

"There are sunspots, of course, but there are also solar flare-ups, associated with the spots and identified first by the English observer Richard Carrington. A glare can last for 30 minutes and emit a thousand times more energy than whole the Sun from an area of only ten thousandth of its total surface ".

From the undeniable affirmation of E. N. Parker we must accept that the sun constantly loses an enormous amount of mass that is transformed into energy of radiation.

At present, according to Parker, the sun radiates energy at the rate of 10^{33} ergs per second. It is obvious that a younger sun must have emitted more energy since the energy gradient should have been higher at the beginning of its life. If we assume that the loss was constant we would have to accept that in 10,000 million years the sun's radiation would have been: 3.1536×10^{50} ergs.

The commented data allow us to compare the current mass of the sun and infer, empirically, the loss of mass suffered by our star through different forms such as: birth of planets, radiation, solar wind, glares and others.

Considering that 8.99×10^{23} ergs are equivalent to 1 kilo-mass, it is estimated that the Sun's mass loss due to radiation, in its life of 10.000 million years, has been at least:

$M = 3.50789766 \times 10^{26}$ kg.

d) Solar Mass Lost in 10,000 Million Years

We have determined that the Sun's mass loss is the sum of:

Solar wind	1.188×10^{27}	Kg
Planets' birth	2.641×10^{27}	Kg
Radiation ...	$0.350789766 \times 10^{27}$	Kg
In 10×10^{10} YEARS....................	2.6563878×10^{27}	Kg.

In order to elaborate this work certain assumptions were established

- Assumption 1: The solar planets had born from sun. The mass of planetary satellites was not considered in this paper. Ex: The Moon

- Assumption 2: The losses due to specific phenomena of radiation such as glow and other have not been considered.

- Arbitrary Assumption: Energy needed to put each planet into orbit has been decided equivalent to one hundred times the planetary mass, but it was not considered as a Sun's loss.

e) Impact of the Solar Mass Decrease over its Planets

As we have seen in the previous paragraphs; during the 10,000 million years that the scientific community assigns to the sun, our star has lost at least 2.656388×10^{27} kg of its original mass.

The inescapable consequence is that the force of attraction exerted by the Sun on the planets has decreased in that proportion and, therefore, the distance between Sun and planets has increased or other parameters of the dynamic equilibrium have been varied, such as planets orbital velocity, its speed of rotation, others.

GRAPHIC # 6
THE DYNAMIC BALANCE OR UNIVERSAL EXPANSION
SUN EXPULSED PLANETS FROM ITSELF, SUCCESIVELY

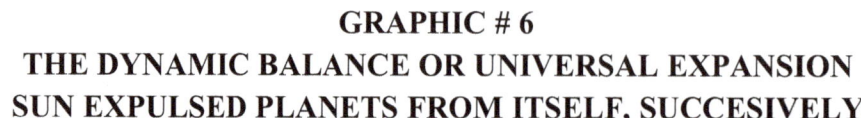

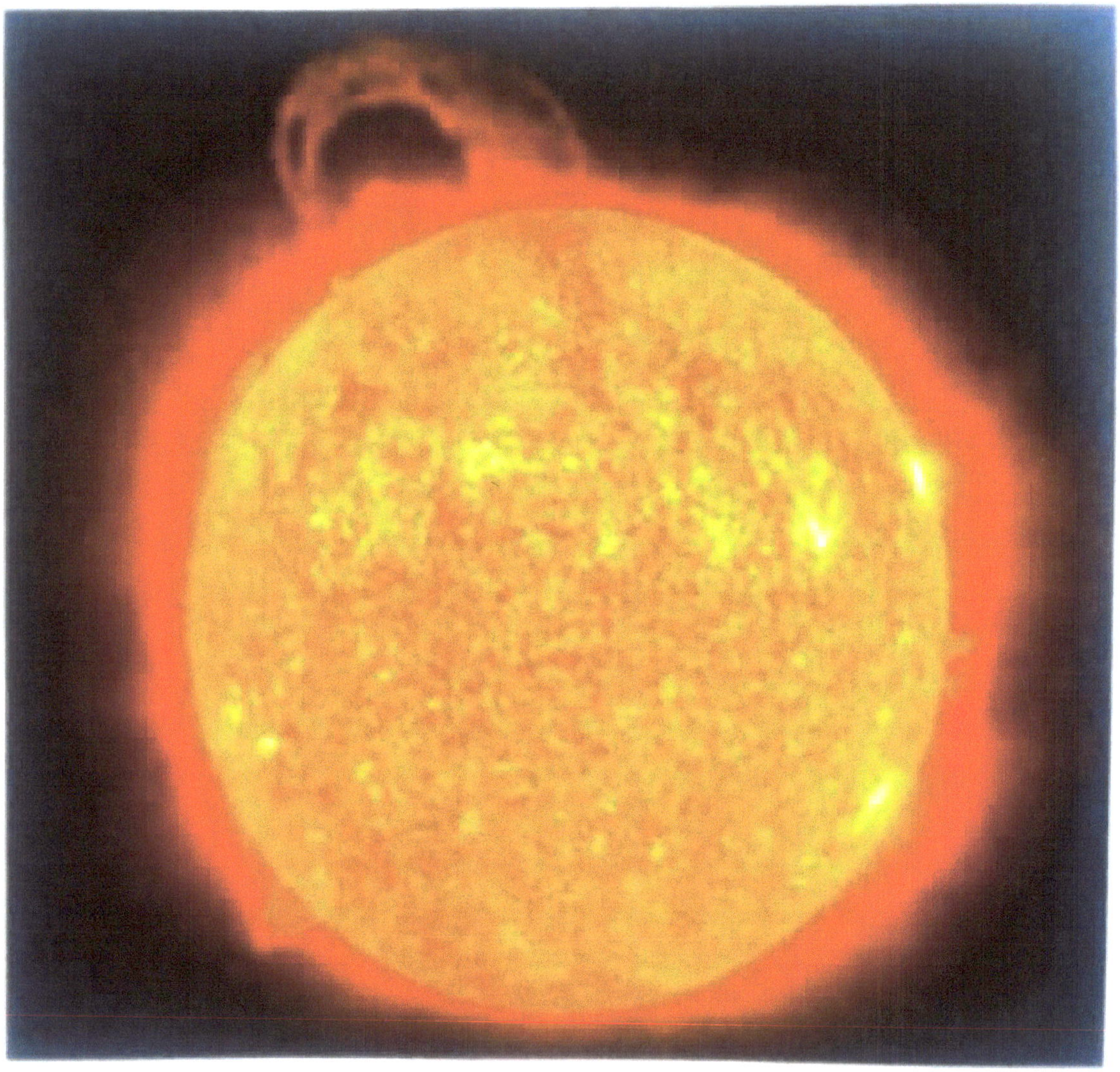

The diagrams on the next pages show the different forms of movement corresponding to: Pure Unidirectional Movement, Pure Rotational Movement and Translation Rotation Combined Movement. The latter is the one that corresponds to the planets revolving around the Sun. It is absolutely clear that the planets have additional movements that are not included in the shown schemes.

9) ROTATION OF AN EME

A planet that orbits a given orbit around the center of the system while rotating on itself (any of the planets of the solar system) had to begin its rotation on itself in a moment and for a given reason. The principles of "rotational dynamics" can help us understand the process.

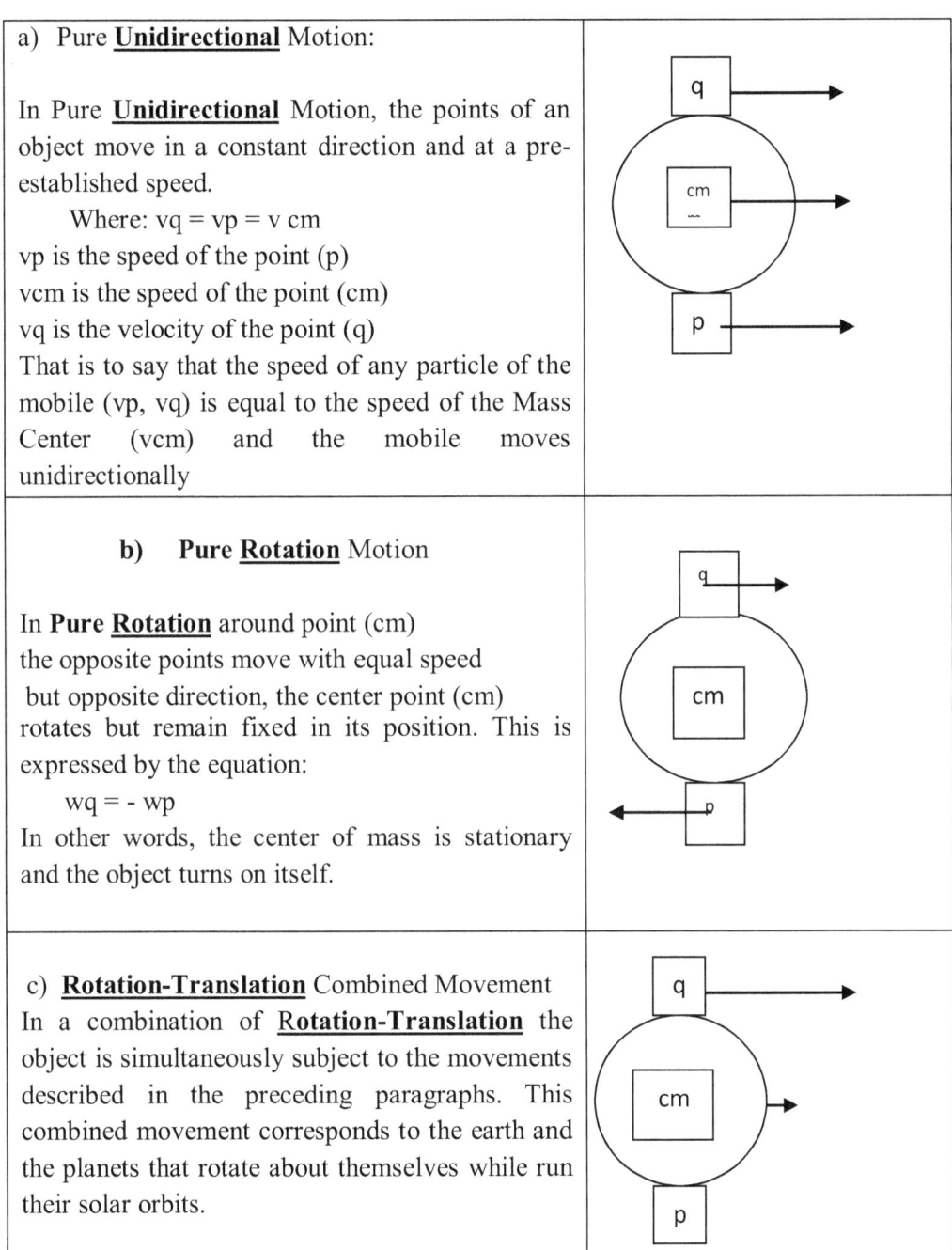

a) Pure **Unidirectional** Motion: In Pure **Unidirectional** Motion, the points of an object move in a constant direction and at a pre-established speed.   Where: vq = vp = v cm vp is the speed of the point (p) vcm is the speed of the point (cm) vq is the velocity of the point (q) That is to say that the speed of any particle of the mobile (vp, vq) is equal to the speed of the Mass Center (vcm) and the mobile moves unidirectionally
b)   Pure **Rotation** Motion In **Pure Rotation** around point (cm) the opposite points move with equal speed but opposite direction, the center point (cm) rotates but remain fixed in its position. This is expressed by the equation: wq = - wp In other words, the center of mass is stationary and the object turns on itself.
c) **Rotation-Translation** Combined Movement In a combination of **Rotation-Translation** the object is simultaneously subject to the movements described in the preceding paragraphs. This combined movement corresponds to the earth and the planets that rotate about themselves while run their solar orbits.

Accordingly, each planet began its movement of translation around the Sun as a mass of planetary dimension was expelled from the Sun and trapped in orbit. That mass had a nucleus large enough to generate gravity and concentrate it mass on a spherical body.

When that mass it was expelled and while moving away from the center of the sun, the movement of the new planets mass was **Pure Translation**, it did not rotate.

When the new planet was trapped in orbit, the conditions of **Pure Translation** around the Sun were modified by the introduction of an imbalance or speeds difference between the symmetrically opposite points of the new planet; that started the rotation.

When the planet rotates as it travels its orbit, the points (q), (c) and (p) describe circumferences of different radius in equal time; consequently, their speeds are different. (See diagram: **Rotation-Translation** Combined Movement).

In these conditions, to movement of "**Pure Translation**", the planet incorporates the "**Pure Rotation**" and from that moment moves, in its solar orbit, in the condition of " **Rotation-Translation** Combined Movement." The new planet thus began the spin over itself. However, it is to be considered that the process took a long time.

10) SOLAR SYSTEM'S DYNAMICS

Whatever their dimension; all 'Mass -Energy' Entities (EME), of the universal space, are subject to the laws of Universal Gravitation. These laws include, particularly, those of the movement.

a) Planets Movement in the Solar System

The planets of the solar system are in permanent compound motion; from the analysis of those movements we can extract the following components.

- The entire solar system is moving with VIA LACTEA.
- The entire solar system follows the expansive spiral movement of the VIA LACTEA.
- In response to universal gravitation the VIA LACTEA is in the process of expansion, consequently the solar system is moving away from the galactic center.
- Each of the solar system planets revolves around the center of the system (the Sun) according to an elliptical trajectory called orbit.
- The speed of the planets varies according to certain proportionality. In their elliptical orbits, the planets use the same time when they travel through elliptical arcs that subtend sectors of equal areas. (See Graphic # 7: DATA OF MOTION OF SOLAR PLANETS)

The table in Graphic # 7 combines some of the characteristics that correspond to solar system's planets; from those characteristics it follows that:

- For each planet, the relationship between SIDERAL PERIOD and ROTATION PERIOD is characteristic.
- The young planets MERCURY and VENUS slowly rotate on themselves; for that, their days are "almost" equivalent to their years.
- From the EARTH onwards, the rotation times become smaller, without apparent relation to their diameters and / or masses.
- The planetary axes declination is similar for MERCURY, EARTH, MARS, SATURN and NEPTUNE with values ranging from 23 ° 27 'to 28 ° 48'; for VENUS and JUPITER are close to 3 °, the URANO rotation is atypical; for the scientific world PLUTON has ceased to be a planet.
- The temporal dimension of a planetary year is given when the planet completes a total orbit.
- Each planet revolves around its own axis. The axis extremes define the planet poles.
- A turn over itself establishes the day's temporal dimension, for that planet.

b) Amount of Movement

In the Solar System the movements of the planets (rotation, translation and declination of the axis) are intimately linked by the principle of "Movement Amount Conservation" that was enunciated by Newton in his second law, with its equation:

$P = m.v$

- WHERE:
- P = Amount of movement
- m = Planetary Mass
- v = Planet total speed

According to this principle, it is found that:

- *If, the sum or resultant of the external forces that act over a body is zero, the body's **amount of movement** is constant.*
- The attraction force exerted by the Sun over each planet and vice versa, is function of the masses; any variation in the magnitude of those masses becomes an "external force" that will act over all the system and over each planet. According with the principle of:
- "Movement Amount Conservation", any mass change produces a movement amount variation in the system and in each one of its components.

- Inside the Solar System; the movements of rotation, translation and planet axes inclination are intimately linked; this is shown in the Table of Graphic # 7, stating that:
- The total speed of the planets has decreasing values, from MERCURY with 48 km / sec to PLUTON with 4.9 km / sec.

GRAPHUIC # 7
DYNAMIC BALANCE OR UNIVERSAL EXPANSION
SOLAR'S PLANETS, DATA OF ROTATION AND ORBITAL PERIOD

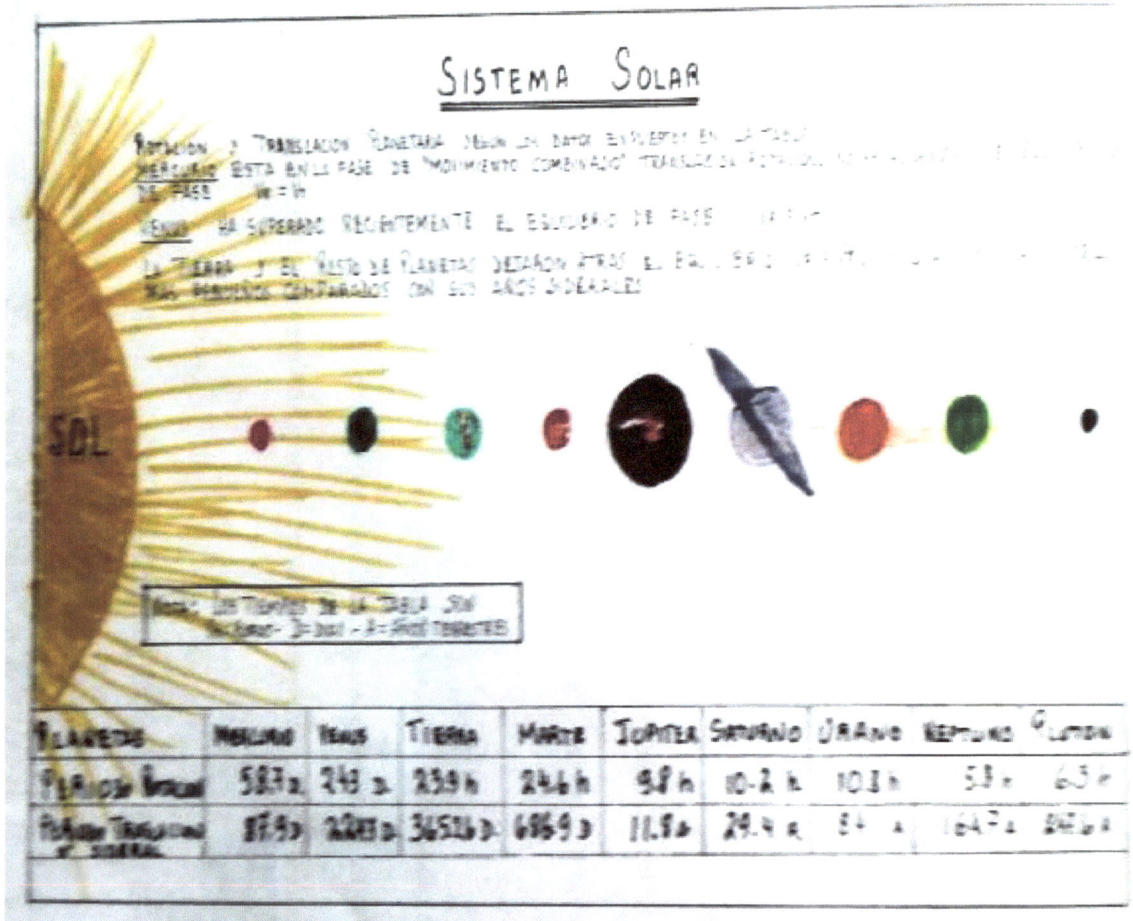

For 'Rotation Period' data of Uranus, Neptune and Mercury; refer to GRAPHIC # 3

c) Effects of Solar Mass Lose

The constant lose of solar mass produced solar attraction force decrement over the planets; as a consequence the solar system is expanding constantly. The percentage of solar mass losing is tiny, in human time, but it affects some planets characteristics as: distance to the Sun, speed of rotation, orbital speed and others.

Whereas, the loss of solar mass is slow and small, the impulse that produces is insufficient to, in human time, alter significantly the planet orbit. However, a large loss of mass can break the dynamic balance and produce hecatomb consequences in each planet.

On the other hand, the variation of distance between planets and Sun is also affected by the constant change of relative position between the planets. This circumstance introduces factors that generate changes in the system's movement amount, but also in each of its components.

From the TERMODINAMICO point of view, it is interesting to indicate that if a planet (like the Earth) always had a "zero" declination, the climate had no changes along time and it was always equal, without our climate seasons. If this planet rotated at a sufficient distance from the Sun, the equator of this planet would have a permanent warm climate, while the poles would be gradually loaded with ice in a process, similar to Earth's Ages of Ice. Was that what happened to the earth in past? At some point in your past the earth's declination was close to zero?

On the other hand, it is evident that a more or less large and rapid inclination's variation of the axis (declination) would produce dynamic forces, large enough, that explain enormous variations of crust's structure, the planetary growth, the emergent material that forms the ocean floor, the progressive distancing of the continents, the enormous superficial force that originated the mountain range systems and many other phenomena.

11) MASS AND ENERGY EMIED BY PLANETS

DYNAMIC BALANCE IN PLANETARY SYSTEMS

The universal attraction has extracted enormous amount of energy and matter from the Sun; but this is also true in the case of planets because they also gives energy and matter to the space, from their birth.

For each planet in its own orbit, the principle of dynamic equilibrium demands certain parameters of: mass, distance, orbital velocity, speed of rotation and others. But the same principle also demands that, if any of these parameters vary, the system must have the capacity to restore the dynamic equilibrium corresponding to the new conditions; otherwise the system would collapse.

It is a fact that the Sun and the planets lose mass permanently, but that loss has not produced the collapse of the system. Consequently, the equilibrium has been restored by changes of other parameters such as orbital velocity, rotation speed, change of planetary declination, distance between planets and Sun. All of which entails changes in physical dimension and density of the system components; Sun and planets.

a) Centrifugal Force VS. Centripetal Force

In a planetary system in dynamic equilibrium, whatever its dimension, the Force of Attraction or centripetal (Fa) is equal to the Centrifugal Force (Fc)

That is to say: (1) Fa = Fc

The components of these forces were studied and defined by Newton.

(2) $$Fa = G \frac{M \cdot m}{r^2}$$

and

(3) $$Fc = m \frac{v^2}{r}$$

WHERE:
M is the current mass of the focus (Sun)
m is the current mass of the satellite (Planet)
G is the universal attraction constant
r is the current distance between masses
v is the current orbital speed of the satellite

Analyzing the equations (2) and (3) it follows that, in case of a decrement in magnitude of system's center Mass (M), the Force of Attraction (Fa) will be proportionally smaller and, in contrast, the Centrifugal Force (Fc) would become greater.

This would lead to the instantaneous imbalance of the system and equation (1) would be transformed into:

(4) Fa < Fc

If the imbalance is not compensated, it would tend to increase indefinitely, producing the collapse of the system by flight (increase in distance) of the planets. We know that this is mechanically impossible because; the moment of inertia (vm) of a rotating body is constant. In other words, when instantaneous imbalance occurs, the satellite decreases its orbital velocity, and increases its speed of rotation, this makes the (Fc) becomes smaller and the equality of (1) is restored; that is:

(5) Fa = Fc

In case of the Sun and its planets, the difference between (M) (solar mass) and (m) (satellite mass) is very large, so the (relatively) negligible losses in (M) are absorbed by humanly imperceptible adjustments of: distance (r) from the satellite to the system center, satellite orbital speed (v), satellite rotational speed and / or changes in declination.

The unobjectionable existence of the solar wind and the energetic radiation of the stars show that the loss of focal mass (M) in any sidereal system is constant. As a result of these factors, the force of attraction of the focus on the components of the system constantly decreases. The natural response of the satellites of the system is to reduce their orbital velocity (v), increase their rotation speed and increase their distance (r) to the focus of the system.

This principle is extended to all the systems formed by a focus, that concentrates a very large percentage of the system mass, and elements that revolve around it; as stellar nebulae or systems that are constantly expanding.

Considering all the above, science can affirm that: in the solar system the energy center (Sun) loses mass-energy constantly, which is proven by:

- Solar Wind
- Energy dissipated in heat the form, energy waves of different nature, others
- Planets born from the Sun

In consequence:

- The expansion of the Solar System begins with the birth of the system.
- The planets were born from the Sun as a result of gigantic energy releases that, in individual events, putted in diverse solar orbit an appropriate magnitude of matter.
- The planets move away from the Sun constantly
- Each planet is, in itself, a system in constant expansion.
- The expansion of the planets is a consequence of their own loss of mass-energy
- The mass releases, or other solar imbalance of any magnitude, are reflected throughout the system.

12) THE SOLAR SYSTEM ... DOES IT EXPAND?

EMPIRICAL APPRECIATION

The updated Solar System data table shown in LAMINA 1 - "DATA TABLE - SOLAR SISTEM", is a reproduction of the one that appears in the article "The Solar System" by Carl Sagan, it summarizes the main characteristics of the planets.

a. The planets density decreases as the distance to the sun increases. Whereas each planet is in itself an EME, the decreasing density in congruence with a greater distance suggests that; the farthest planet has lost a greater percentage of its original mass and consequently, at a greater distance from the sun the age of the planet is higher.

b. Considering that the number of satellites is greater if the planet is larger or its distance to the Sun is greater or both, it can be inferred that there is a law of proportionality that relates the number of satellites with the size of the planet and/or its distance to the Sun and/or its age.

c. The planets orbital speed decreases as the distance to the Sun increases. This phenomenon is related to the principle of "AMOUNT OF MOTION" in the sense that: when the force of attraction (centripetal) decreases, as a result of the Sun mass lose, the planet compensates its "angular momentum" variation decreasing its "orbital speed" while increasing the distance separating it from the center of the system.

d. The decrease in central attraction or the imbalances that occur in an energy system, make the system asymmetric. That affects the equation of attraction. Therefore it can be said that the distance, from the planets to the Sun, increases as an unavoidable consequence of the mass loss suffered by the Sun for different reasons: radiation, solar wind, birth of new planets and others.

e. From the value of planets mass it is inferred that the solar system has gone through different periods of asymmetry.

f. At the initial moment with the mass still concentrated and intact, the solar system was close to thermo spherical symmetry with its maximum gravitational attraction power.

g. The first asymmetry must have occurred within the planetary focus (Sun), the product of that asymmetry was the birth of the first planet.

h. Successive asymmetries gave rise to other planets.

CONCLUSION

- The balanced Universal Expansion is consequence of the Universal Gravitation.
- All the Energy Mass Entities (EME) are subject to Universal Gravitation whatever their dimension or energy state.
- The Dynamic Energy Balance that harmonizes the relationship between EMEs and their environment governs and controls Mass-Energy transfers.
- Any harmonic alteration of the Dynamic Equilibrium of any EME, will affect it in magnitude proportional to the magnitude of the alteration.
- Inharmonic alterations can generate uncontrollable reactions, because they can generate chains of external factors, for which the EME may not be prepared.

Today on earth, governments, companies, armed forces and terrorist groups have the capacity to produce inharmonious alterations; that is why I repeat what I say decades before:

"Can human being overcome ADAM, his mythical father? Will he keep his Paradise?

OUR ONLY ONE EARTH

EARTH

AND

THE DYNAMIC BALANCE

EARTH, DOES IT EXPAND IN A DYNAMIC BALANCE?

Author: Luis Javier Artieda Carpio

BOOK TWO

EARTH AND THE DYNAMIC BALANCE

EARTH, DOES IT EXPAND IN A DYNAMIC BALANCE?

Since the last decade of the previous millennium, humanity is becoming aware of the ecological deterioration that affects our planet; global warming, the great spirals of plastic garbage that grow irrepressible in the oceans and exterminate the aquatic life without apparent solution, the "accident" of the platform of BP (British Petroleum) that in the gulf of Mexico discovered dangerous flanks related to the deep drilling and the so-called "hydraulic fracking" under oceans and continents. These facts are part of a rampant reality that today coexists with the human being.

This rampant reality shows that the rights of humanity are increasingly affected by the free action of companies that, supported by ad-hoc laws, maintain their dangerous operational practices in the limbo of confidentiality and the exclusive operational secrecy with consent of the governments. That is why, today 2017, I think it is pertinent to reproduce what I said in 1987.

1987

Equilibrium or Earth Diameter Expansion"

In Lima at a university conference I said:

"I believe in the natural process of balanced the terrestrial sphere expansion. I do not know if others, anywhere, have written about this matter, I just hope that the idea is evaluated and gets the contribution of clear and educated minds.

As far as I'm concerned, I'll try to deepen this topic in future.

I believe that the energy loss through the crust and the exhaustive measurement of this radiation should be studied and explained in order to establish the relationship between the earth internal geo-thermal processes with that of terrestrial expansion.

The difficulty in establishing certainty about these or similar problems teaches us that this is a field of great uncertainty; however, I firmly believe that defining the matter of the Earth's core as: "mass in the process of radioactive expansion" gives the geodynamic energy required for expansion.

Luis Javier Artieda Carpio

1) HYPOTHESIS: THE EARTH EXPANDS DYNAMIC AND BALANCEDLY

The hypothesis of: "Drift of the continents", "Expansion of the Earth Diameter" and "Tectonic Plates" have been discussed throughout the 20th century"

This work aims to demonstrate that the "Terrestrial Diameter Expansion" Hypothesis corrects some inconsistencies of the "Continental Drift" and "Tectonic Plates" hypotheses, for which it will review important phenomena of terrestrial geology and will propose explanations congruent with the "expansion" "

In this sense, the author states that:

Being demonstrated (by dating samples collected from the ocean floor) that the first rip of the crust gave rise to the Pacific Ocean, it is congruent accept that this enormous geotectonic effort initiated a phenomenological chain whose links include:

The primitive system of mountain ranges in the original America, was formed by the Andean mountain range, Rocky mountains of North America, Sierra Madre in Mexico and others in the terrestrial area that later will be known as Central America. All these chains were connected in succession through the primitive continent.

The separation process between Alaska and the Chukotka Peninsula gave rise, on the Asian side, to the Koriakor and Kolima mountains, not yet divided; the Seas of Japan, Yellow and of the South China that were not more than small cracks in process of widening.

The Indochina Peninsula was together to India, Sumatra, Jaba, Borneo, the Celebes and Australia, near Africa and Antarctica, occupied almost the entire Indian Ocean, and the widening of that ocean projected them to their current geographical positions.

America was together with Africa, which still was separated from Europe for a thin crack that would become, through the millennia, in the Mediterranean Sea.

What later would be "La Patagonia" was together with the Cape of Good Hope and welded to the Antarctic Continent by the actual outer edge of the Antarctic Peninsula.

This sui generis panorama was part of the primitive terrestrial crust, in which large and flat continental plains did not yet present the accidents of today, and most of it was submerged in an apparently larger ocean than the current one, with a similar volume of water to our current oceans, it covered a terrestrial sphere of perhaps 10,000 kilometers in diameter.

As already mentioned, geological studies carried out since the nineteen forties assign the longest antiquity to the Pacific Ocean (Mesozoic) trench. The birth of the Atlantic dorsal is posterior. The Indian Ocean is contemporary to the Pacific but its process is different.

As it expanded, the Earth increased its spherical surface over which it carried the original continental pieces projecting them in radial form.

The difference in curvature between the base sphere and the thin, but rigid, hull of the crust caused superficial efforts of action and reaction reflected in the formation of oceans, ridges, insular arcs, marine pits, river basins, depressions, interior seas, lakes, gulfs, peninsulas and much more.

On the other hand, the existence of bands with opposite magnetism, parallel to the oceanic ridges, generated the Magnetic Polarity Inversion hypothesis of the Earth Poles every certain number of millions years. This work proposes, later, an alternative explanation.

2) ANCIENT CULTURES CALENDARS

In 1970 I became interested in these subjects, and at the same time I learned that the Maya Calendar testified; the temporal dimension of the Earth year was only 360 days. At first I didn't pay attention to this detail.

There is no reason to deny this claim, the Maya people have left countless and indisputable samples of precision, astronomical knowledge, scientific interpretation capacity and technological advance.

On the other hand, many other ancestral calendars coincide in that there were changes in the temporal dimension of the solar year and the lunar year. Some of these calendars accuse a solar year of 360 days. To that temporal dimension are applied corrections for to reach the dimension of 365.26 days of the current year.

That change suggests that the orbital speed and / or the dimension of the Earth and Moon orbits have changed. It also states that the Earth-Moon system would have moved away from the center of the solar system (the Sun)

Considering the changes that Maya, Egyptian, Indian, Chinese and others accused in their calendars, I wonder:

Do those changes reflect some Earth's orbital velocity change? In what was the lost kinetic energy transformed in that change of speed? That is to say: How was it that the 360-day land year, registered by them, became our bath of 365.26 days?

Does it mean a change in the kinetic equilibrium? In other words, did the deceleration of the Earth imply any decrement in terrestrial Kinetic Energy?

This obeys the general formula:

$Ec = ½(2/5 m.r^2) w^2 + 1/2\ m.v^2 +$ others

PREMISE: Energy is not created or destroyed, it only transforms!

Kinetic Energy of an EME and the Mayan Calendar

Every EME is subject to different forms of Kinetic Energy that added; result in EME's Energy total. This is expressed in the following formula

$Ec_T = Ec_r + Ec_t + others$

Where:
Ec_T = Total Kinetic Energy
Ec_r = Rotation kinetic energy
Ec_t = kinetic energy of orbital movement

This relationship is represented in the following general formula:

$Ec_T = ½ (2 / 5m.r^2) w^2 + 1/2 m.v^2 + others$

For this study, we will only take into account what corresponds to:

$Ec_r = ½ (2 / 5m.r^2) w^2$ (Kinetic rotation energy)

and

$Ec_t = 1/2 m.v^2$ (Kinetic Energy of planet's Orbital movement)

and its equivalent

$1/2 m.v^2 = 1/2 m (S / t)^2$

Where

'S' is the length of the orbit
't' is the travel time of the orbit.

However, remember that ancient cultures calendars such as the Maya recorded an annual time of only 360 days; this tells us that the time 't' increased to $(t + \partial t)$ which makes the equation of 'Ec_t' change to :

$Ec_t = 1/2 m [S / (t + \partial t)]^2$

If the data of the ancestral Calendars is true, the Earth system had to respond by changing in some way the parameters of Ec_t, Ec_r and others; but, preserving the value of Ec_T to fulfill the premise: Energy is not created or destroyed, it is only transformed!

In other words, the change in Ec_t indicates an increment in 'S' (orbit length), but 't' (orbit travel's time) also increased according to:

$Ec_t = ½ m [(S + ∂s) / (t + ∂t)]^2$

In this way the energy equivalence would have remained.

a) If on the contrary, Ec_t changed with

$(S / t)^2 < [S / (t + ∂t)]^2$

The entire system will be affected by the decrease in orbital speed and other changes must have occurred to compensate.

In conclusion, the variation of (t) will also bring about compensatory changes in other variables of Ec_r equation (rotation's Kinetic Energy).

$Ec_r = ½ (2 / 5 m.r^2) w^2$ that would change to

$Ec_r = ½ (2 / 5 m.r^2) (w + ∂w)^2$

The changes in Ec_r and Ec_t makes Ec_T to remain intact.

As we know our Sun, loses mass and energy every second in very large quantity and, through catastrophic events, a no quantifiable losses that affects and had affected our planet.

If the ancient Maya Calendar's observation was true, we can conclude that:

- The yearly increment of 5.26 days in (t) (travel orbit time) must have produced a significant change of the Ec_T (Total Kinetic Energy)
- That is, at the time of the Maya measurements and other cultures, the translation speed was higher since the planet traveled its orbit (S) in a time (t) of 360 days, today it does so in 365.26 days.
- By definition, Ec (Kinetic Energy) is constant, in consequence any change in time of a phenomenon, implies that other equation parameters will changed to compensate.
- Considering that the ancient cultures observed and recorded a significant variation in the Earth orbital speed, we must ask: What caused the change in the orbit time (t)? What other parameter has changed because of that variation?
    - The orbit length (S) seems to have changed, but just as time (t) is "dependent variable", consequently (t) and (S) have no capacity to vary by themselves.
    - The distance (r) from the Earth to the Sun would have varied, but it is a "dependent variable", consequently (r) does not have the capacity to vary by itself.
    - The orbital speed (v) of the Earth would have varied but it is also a dependent variable, consequently (v) it has no capacity to vary by itself.
    - The angular or rotational speed (w) of the Earth would have varied but it is also dependent variable, consequently (w) it does not have the capacity to vary by itself.

- o Of all the variables described, the Earth mass (m) and the Sun mass (M) vary constantly in response to Universal Gravitation. Consequently (m) and (M) are the only independent variables of solar system Kinetic Energy.
- If we accept the increase of 5.26 days in the year we must accept that the Earth lost mass in significant amount; the logical consequence is: the ancient calendars give certainty that the Earth's mass variation can be observed and maybe measured in human lapses.
- Another logical consequence is: thanks to the ancient calendars (like the one of the Maya) we could affirm that the celestial bodies lose mass and that loss is reflected in expansion.
- Another logical consequence is: thanks to the ancient calendars (like the one of the Maya) we could affirm that the celestial bodies lose mass and that loss is reflected in expansion.
- By similarity, we can infer that: every celestial body will expand as a result of its loss of mass and energy and, as the Earth is a celestial body that loses mass constantly, THE EARTH EXPANDS.

3) THE EARTH

For purposes of this work part, I will consider the planet Earth with features of 'Mass-Energy Entity' (EME) that was already defined in 'Book 1' (THE DYNAMIC BALANCE OR UNIVERSAL EXPANSION).

Earth, like all planets, is an EME in process of reducing its mass and energy and expanding its volume.

The primitive solar system energy center; it had expanded giving rise to different proto-planets, one of which was Earth.

Through time, the proto-planet Earth radiated energy and matter, at the expense of its mass and, little by little, it became a planet.

Due to the constant loss of energy, the outer layer of the Proto Planet solidified; forming the original rigid crust. From that moment on, any inharmonic accumulation of energy was released through thermo-tectonic processes of variable magnitude. In the consolidation stage, the crust temperature should have been high, plastic texture and in some parts elastic. Chemical changes, such as oxidation processes and others, transformed this plasticity into rigidity and wind and hydraulic erosion began.

a) The Terrestrial Expansion Motor

Starting from the assumption that the Earth has a life of 4,500 million years and, considering that crust's oldest rocks age is approximately 3,500 million years, we assume that the pre-geological time of the Earth was 1,000 million years.

In its pre-geological time the Earth behaved like a micro-star; so, the energy loss was very great. The 'small' Earth dimension allowed a relatively quick energy loss compared to larger bodies such as the Sun. Through its pre-geological time, from the interior of this proto-planet in cooling process, low-density material emerged that has covered the surface of the sphere and formed the primitive rigid crust. However, the interior of the balloon continued to emit enough energy to keep the crust unstable and at high temperature.

If we accepts that the energy released is produced in radioactive processes generated by of heavy and dense elements, when are closest to the center of the Earth, it would be logical to suppose that the center of the terrestrial sphere is formed by a very dense nucleus, in which the atomic limits had no born yet, and the earth nucleus is formed of a mass of "nuclear plasma" of very high energy, in dynamic equilibrium.

The transformation of nuclear plasma to the elemental forms give birth the atom. This process generates radioactive energy release; logically the volume increases and the density decreases.

We have seen that the energy released from the planet's inside follows a complicated process, that it is produced in "thermo-energetic dynamic equilibrium". This process happen, necessarily, connected with the variability of the balloon dimension. Consequently, it is necessary to accept that the planet expansion is a consequence of mass and energy loss.

From the preceding analysis we can conclude that:

- At the end of its proto-planet period Earth was a of smaller diameter sphere than in this days.
- Initially, the crust was formed by all continental pieces together, without significant deformation.
- As nuclear plasma increases in volume during its "condensation" to atomic forms, it produces the following phenomena: energy escapes to exterior space, pressure increase inside the sphere, sphere is expanded and it produces crust deformations.
- The expansion cracked the crust and allowed material flowed from the inside to surface. The emerging material filled the cracks, joins the rest of the crust and formed part of the ocean floor. This happen currently when magma flows by rifts or volcanoes, through continental or submarine crust.
- The constant expansion process, leads to a permanent state of dynamic equilibrium, in constant changing.

- The expanding terrestrial dynamic-balancing process obeys to the loss of mass and energy, but it is affected by several factors such as: changes in rotation, changes in its orbit travel and many others.

4) CENTRIFUGAL FORCE VS. CENTRIPETAL FORCE

In Book 1, "The Dynamic Balance" we said that: In a planetary system, whatever its dimension, the Centripetal Force (Fa) is equal to the Centrifugal Force (Fc), from instant to instant.

That is:

(1) Fa = Fc

In which:

(2) Fa = G. M. m / r²

and

(3) Fc = = m v² / r

Where:
M = terrestrial mass
m = mass of each of the elements of the bark considered in isolation.
G = the universal attraction constant
r = the distance between the Earth center and a bark part, considered in isolation.
v = Earth's equator's rotation speed.

Analyzing equations (2) and (3) it follows that in case of a decrease of Earth's mass (∂M) by mass and/or energy radiation, the Force of Attraction (Fa) will be proportionately smaller and, by contrast, centrifugal force (Fc) would be comparatively greater, as stated in equation (5)

(4) $G\frac{Mm}{r^2} = m\frac{v^2}{r}$ - If M decreases –

(5) $G\frac{(M-\partial M)m}{r^2} < m\frac{v^2}{r}$

This leads to the instantaneous imbalance of equation (1) that becomes:

(6)   Fa < Fc

If the imbalance is not compensated, it tends to increase indefinitely leading to the planet's collapse; but we know that this is mechanically impossible because the Moment of Inertia (vm) of a rotating body is constant. In other words, when instantaneous imbalance occurs, the planet increases its diameter; this makes the relationship [v2 / (r + ∂r)] decrease. This decrease affects the (Fc) making it smaller and the equality of (1) is restored; that is to say, when increasing the diameter the Dynamic Balance is restored.

(5) $G\frac{(M-\partial M)m}{(r+\partial r)^2} < m\frac{v^2}{r+\partial r}$

This restores equality in (1):

(1) Fa = Fc

Considering that the difference between M (planetary mass) and ∂M (the loss of planetary mass) is immense; those small losses are absorbed with (imperceptible) adjustments of: planetary diameter (r), rotation speed, others not yet identified.

The occurrence of earthquakes, hurricanes, cyclones, volcanic eruptions and other thermodynamic phenomena show that the mass (M) loss is constant; therefore, the force of internal attraction (gravity) decreases and, by contrast, the force of external attraction (centrifugal) increases constantly. Consequently, the planet increases its diameter constantly. Obviously, the density decreases and other parameters such as translation and rotation speeds, rotation times and others also vary.

From this it follows that

- Every planet is a Mass-Energy Entity (EME) in constant expansion.
- The expansion of the planets is also result of loss of their own mass-energy.

5) RADIOS OF SPHERES THAT GROW

Starting from the assumption that oldest rocks of the current terrestrial surface are pieces of the primitive crust and as old as it; it is licit to suppose that everything that existed before that rocks, corresponded to a terrestrial sphere without consolidated crust. (See Graphic # 1 EARTH EXPANSION PROCESS) (Graphic # 1 is from 1987, unpublished).

At the risk of assuming major errors, we accept that the Paleozoic terrestrial crust was composed by the sum of the surfaces of the current continents (148 940 000 km²) added to the continental platform, and increased by a percentage due to the contraction of the original crust during its cooling and subduction processes'.

Under these conditions, the diameter of the Earth would have been somewhat greater than ten thousand kilometers (10,000 = Km) and its total surface area, some three hundred and fourteen million and one hundred and sixty thousand square kilometers (314,160,000 km2).

In these conditions we can affirm that, in the Paleozoic the earth had an almost rigid crust with curvature corresponding to the diameter of that moment (10,000 Km.).

With the Mass-Energy loss, the Earth increased in volume, and consequently the surface of the sphere increased in direct relation to increment of diameter.

As the terrestrial sphere expanded, the primitive inelastic and rigid crust was torn leaving empty spaces between its pieces. The increase in spherical surface was covered by mantle emerging material, which was added to the lithosphere.

If it were possible to see the progressive increase of the terrestrial surface across the ages, magma would emerge between the pieces of the original crust while the pieces of the crust were projected in radial way to occupy a geometrically similar but relatively smaller space in the expanded sphere.

During the terrestrial volumetric expansion process, the emergent material leaves to the outside by different routes but in greater quantity by the suture crest of the oceanic mountain ranges. The basaltic layer of the continents is stiffer and thicker than the one beneath the oceans and is, therefore, more resistant. Consequently the fracture (by simple principle of materials resistance) occurs on the oceanic crest.

But as we shall see later, must accept that the leak of magma that comes out from under the continents is real, as is suggested by the great island arcs (Islas Marianas and similar). Other phenomena that witness the mobility of the primitive crust is the trace left by India in Indic Ocean bottom in its long runoff. However, other primitive bark structures, such as Africa left similar traces and has not been taken into account by the official geology

Considering what has been said we could affirm that some questions remain pending; what caused the crust fragmentation? Will the expansion continue? Will it be through discontinuous events?

We know that, in order produce a significant expansion to occur, the center of the system must lose an important QUANTUM of energy. Considering that the planet, it loses small amounts of energy per time unit, the expansion occurs in slow motion, except when solar catastrophic events accelerate the planet expansion.

On the other hand, the sphere also expands continuously through lesser processes such as magma flow through volcanoes, growth of vegetation in forests and oceans, and other bio-energy processes, leakage of atmospheric gas sucked by the solar wind, and other unknown or not studied processes.

a) First Expansion - First Fracture

The primitive crust was consolidated becoming rigid; this rigidity prevented it from adapting elastically to the basic sphere in growing process; for this reason the crust was irregularly fractioned. From then on the expansions made the irregular fractures more obvious

GRAPHIC # 1
EARTH EXPANSION PROCESS

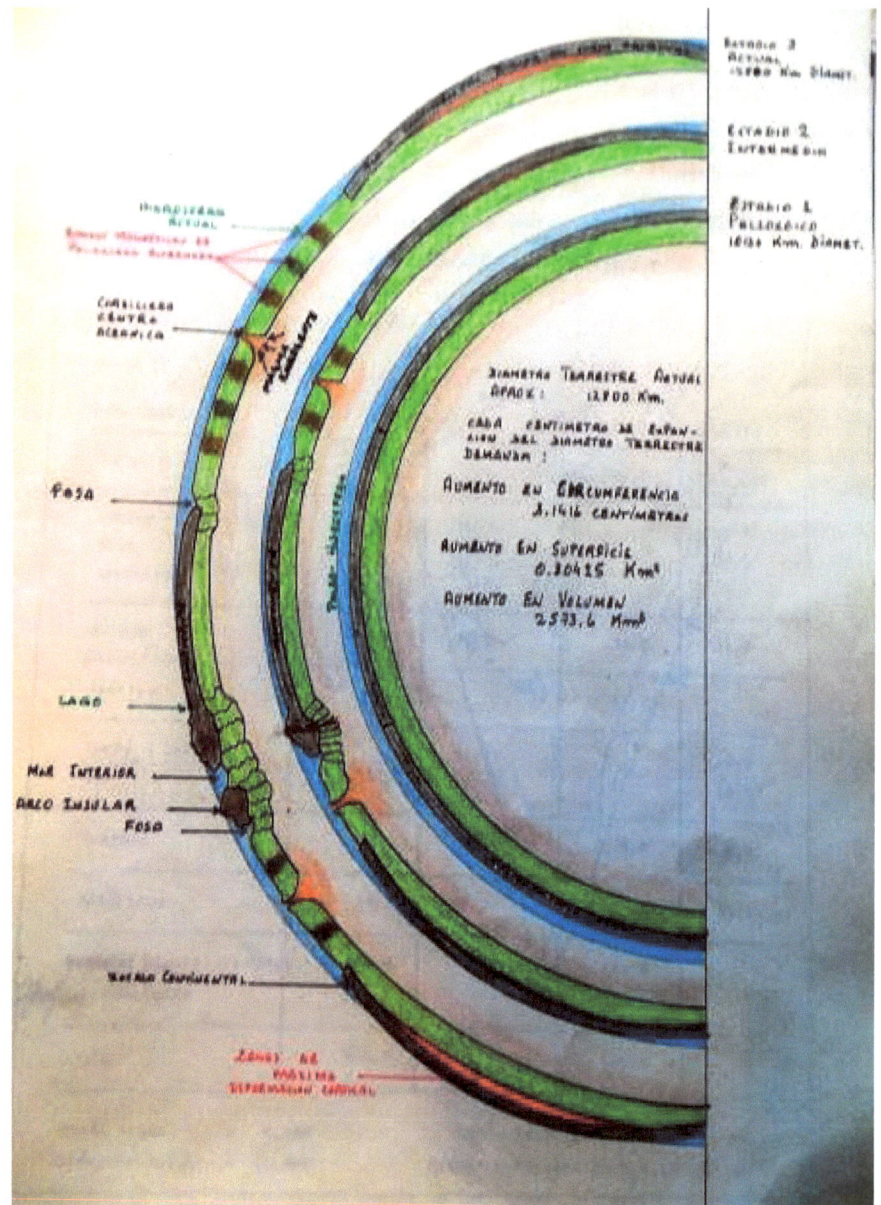

The first large fracture opened the embryo of current Pacific Ocean. With the continuous growth of the base sphere the internal pressure against crust increase. In the first stage the resistance and the stiffness of the bark pieces caused it to retain their smaller diameter spherical shape; however when the internal pressure exceeded their resistance the crust pieces were fractioned. From that moment the pieces of original bark were accommodated to the sphere in constant growth. (See Plates 2, - PALEOZOICO - CENOZOICO - MESOZOICO - LA TIERRA TODAY (The graphics are from 1987, unpublished).

As we see in the graphics; the original crust structure was resting on the mantle surface of terrestrial sphere, where they were subjected to diverse surface compression and traction efforts (One of these important efforts is due to the "Magma-Tectonic Inertial Runoff" that we will see later). Over time the aforementioned efforts produces, in bark pieces: slide off, modification of geometric-radial position, subdivisions and various modifications related to shape, dimension, cortical thickness and others.

The current continents and their "Continental Platforms" are the pieces of the primitive crust modified by successive expansions that are reflected in visible geological deformations. (See Graphics #s 2, 3, 4, 5, 9 and 10)

b) Magma-Tectonic Inertial Runoff

The planets are subjected to all kinds of forces during the expansion process; one of them is generated by the rotational inertia that affects all the planet's structures. (See Graphic # 1 - Earth Expansion Process and Graphic # 9 Magma-Tectonic Inertial Runoff)

In relation to the crust and its parts (continents and others); it must be considered that if, for any reason, the planetary rotation speed changes, the friction between Mantle and Crust will generate a force opposite to the rotational inertia. This phenomenon is exposed in the three stages of Graphic # 9 - Magma-Tectonic Inertial Run off.

Stage 1 of "Earth Expansion Process" chart, the curvatures of base sphere and the curvature of 'crust' are equal while the Angular Velocity (w) is constant and equal for both structures.

Stage 2 of the "Earth Expansion Process" graph represents a moment in the base sphere expansion process in which the continents, already separated, maintain their dimension and a spherical-geometric-geographical position similar to that of Stage 1 but over the base sphere expanded. One of the most important consequences of the base sphere expansion is the planetary angular acceleration; this acceleration is reflected in outcrop of magma on the East edge of the continent "O" and subduction to their West.

Stage 3 of the "Earth Expansion Process" graphic: it represents a subsequent expansion of the base sphere and shows the same phenomena as in the previous phase, although of greater magnitude.

For purposes of the theory of "Dynamic Balance" the phenomenon of "Magma-Tectonic Inertial Slide off" (as we will see later) provides elements of explanation to the existence of the island arcs such as: Solomon archipelago, Guam, Japan and others.

GRAPHIC # 2
EARTH EXPANSION PROCESS
FROM PALEOZOICO TO OUR DAYS

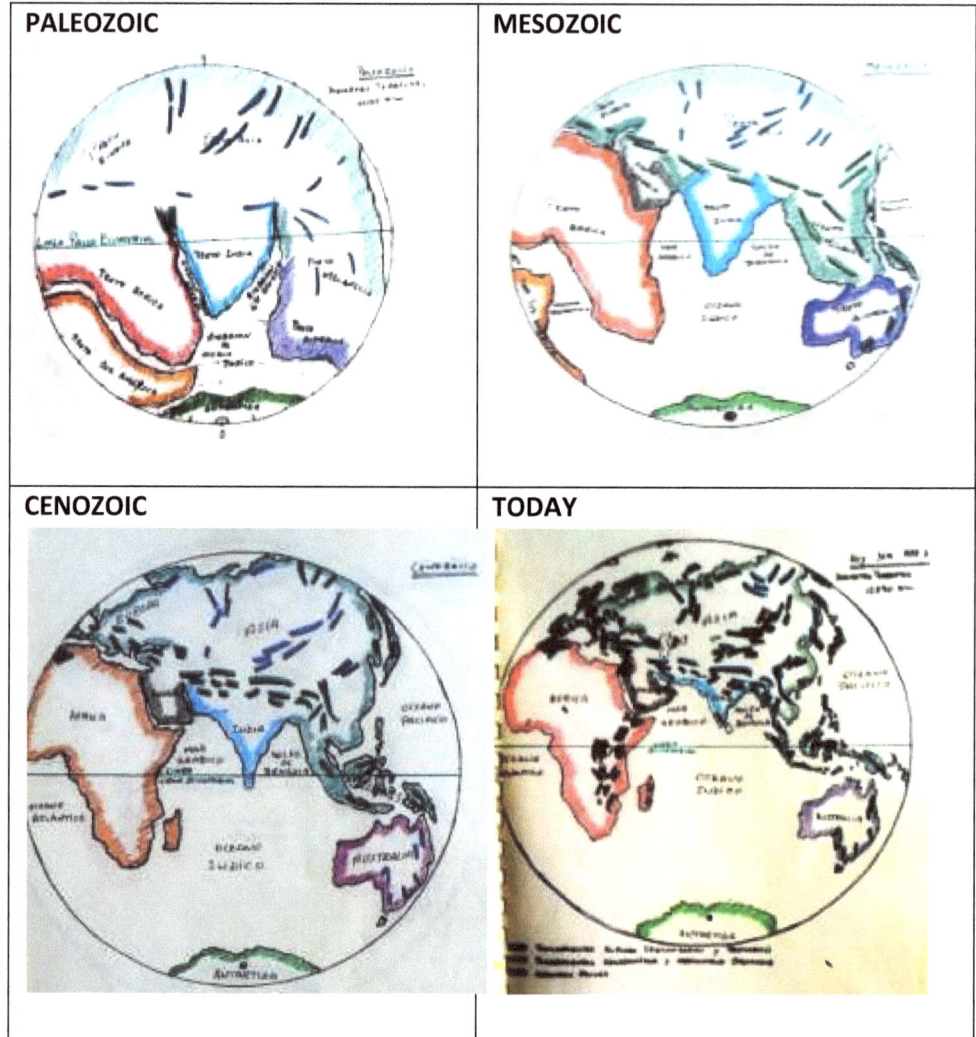

In South America: the South Sandwich Islands. In the Pacific: New Zealand and the Great Barrier Reef in Eastern Australia, The great trench of Peru - Chile and others.

6) CHRONOLOGICAL ATTEMPT (See Graphic # 6 - Lungfish)

How to explain that two living 'terrestrial' species, with common past evolutionary, exists today in the extreme south of two different continents and, halfway between them, in a small volcanic island without connection to any and thousands of kilometers from both continents, exists another one living species undoubtedly related to the two other.

GRAPHIC # 3
EARTH EXPANSION PROCESS
Evolution of Arctic Ocean and
Surrounding Continents

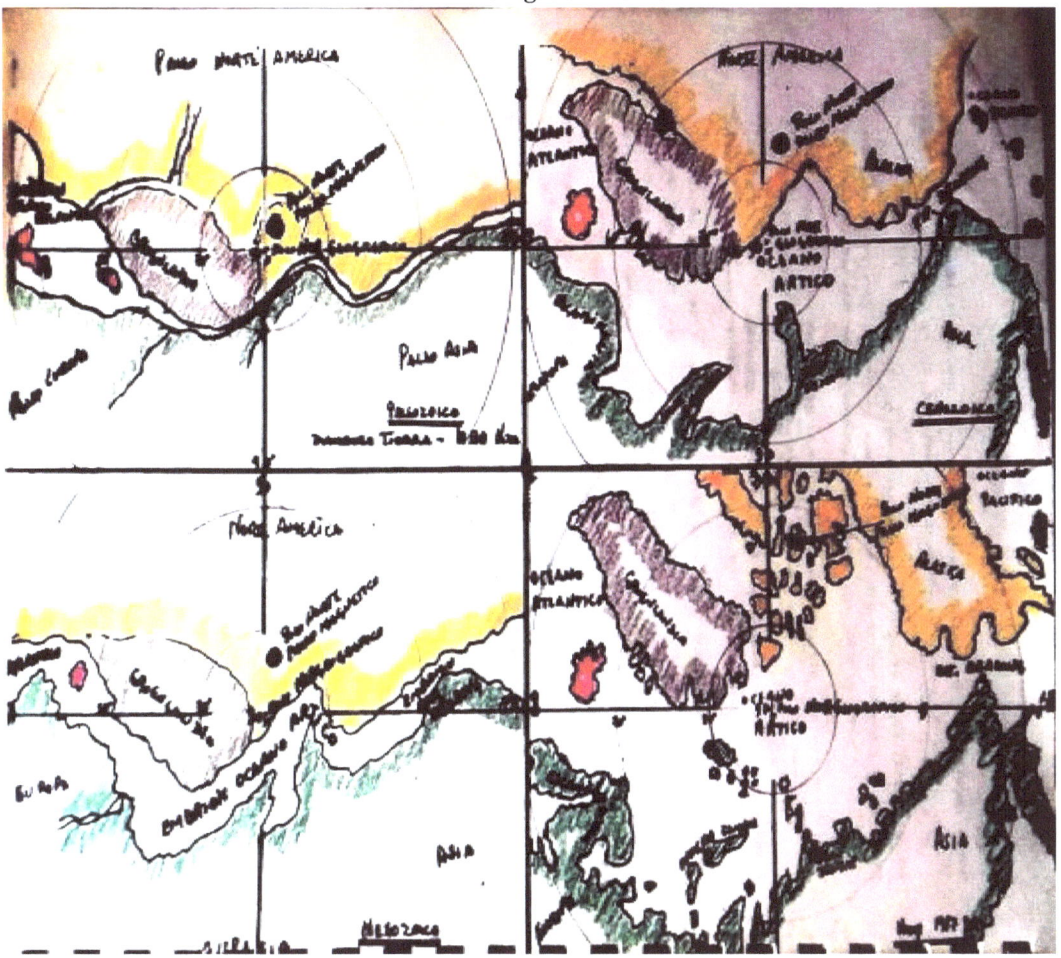

Tristán de Acuña Island was born with the great Atlantic fracture, and it is one of the highest peaks in the South Atlantic Dorsal mountain chain. This island t is almost equidistant from Africa and America (around 3000 km far from each continent). This and other islands on the Great South Atlantic Ridge such as St. Helena, Ascension, Gough, Bouvet, St. Paul's Rocks and other as Azores islands on the Great North Atlantic Ridge would demonstrate that, at the time of the great fracture and outcrop of magma from mantle, America should were almost joint to Europe and Africa following the line the Atlantic dorsal chain.

Those islands, with similar flora and fauna forms to those on both continents; America and Europe or America and Africa shows, that in a remote geological stage the aforementioned islands were connected to both ocean sides. The time period for that connection was long enough to permit the ancestral forms would lived in the three geographical places, should had left offspring and over the millennia should had adapted to its new and changing habitat through distinguishable phenotypic variations. Undoubtedly, ancestral fossils show their common origin.

GRAPHIC # - 4
ANTARCTIC CONTINENT EVOLUTION
SOUTH POLE
FROM PALEOZOIC TO OUR DAYS

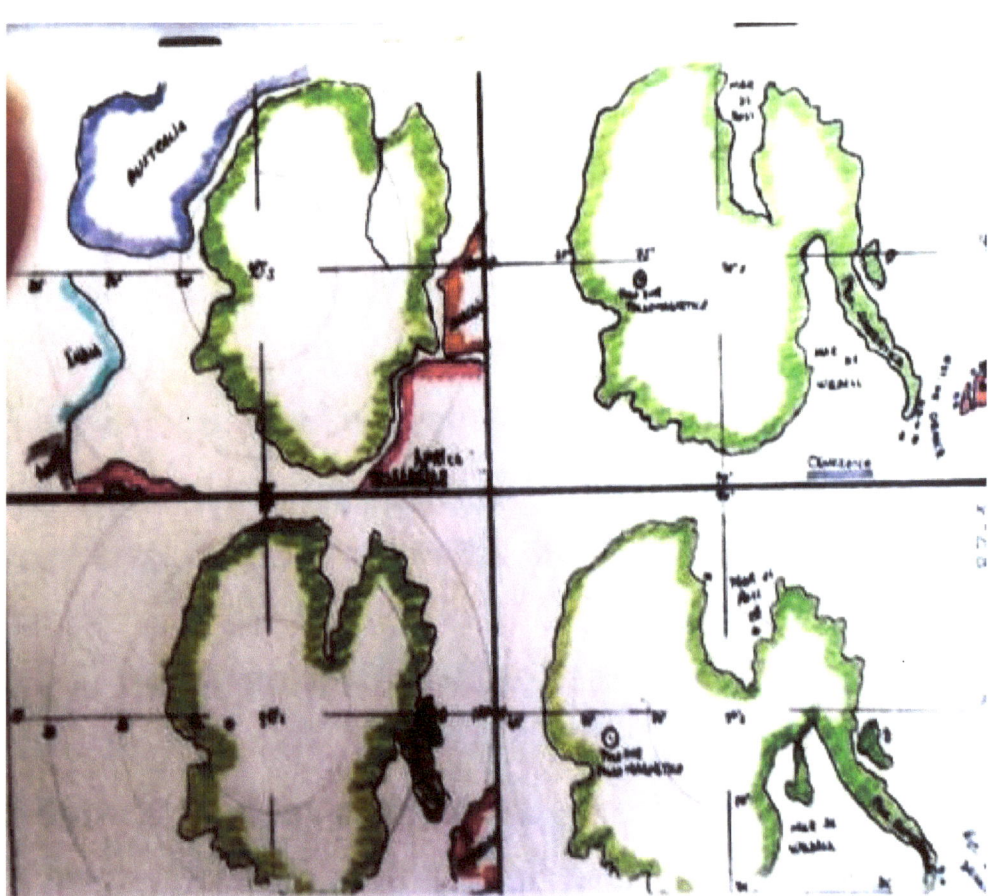

This is confirmed when, similar flora and fauna living forms were find in different evolutionary stages, coming from a common ancestor in geographic places today distant and discontinuous, but irremediably united in remote past.

The Atlantic reality suggests a closeness almost meridian fractionation from pole to pole and, considering the relative closeness between the continental masses, it could mean a shorter antiquity with respect to the Pacific trench. Tests of Paleozoic-dating confirm it.

CLOSEST VOLCANIC ISLANDS: The closest volcanic Islands as Bermuda, Martin Vaz, South Georgia, South Sandwich near south extreme America, o Cabo Verde Islands, Wood, Faeroes, Canarias, and others, near Euro-Africa, were born in later stages than those of the Great Atlantic Dorsal and are related to the nearby continent; it is elementary that their age is less than that corresponding to the islands of the Great Dorsal (Tristan da Cunha and others). (See Graphic # 6).

Sir Gavin De Beer in his "Atlas of Evolution", referring to the discontinuous geographical distribution of the cochineal and other living or fossil organisms (map 22), explains that the winds and ocean currents are responsible for the colonization by these Terrestrial crustaceans and plants, living or extinct, such as Pachijglosa moss in New Zealand and Patagonia.

GRAPHIC # 5

OCEANIC AND TERRESTRIAL MOUNTAIN CHAINS FORMATION

FROM PALEOZOIC TO OUR DAYS

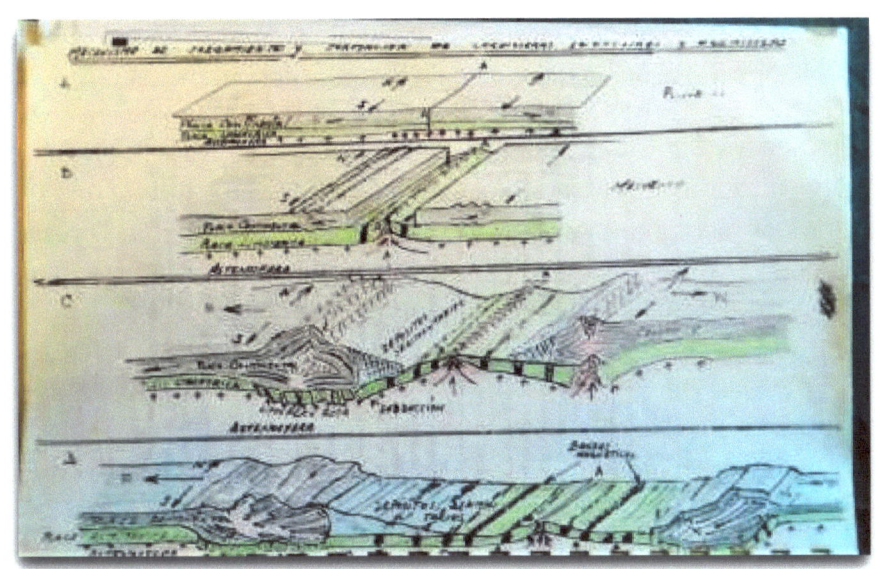

A – PALEOZOIC   B – MEZOSOIC   C – CEONOZOIC   D – TODAY

These living and fossil forms, alluded by De Beer, appeared at the dawn of the planet, at a time when the discontinuous geographical extremes of today were united; when the Earth's crust covered a smaller base sphere. The growth of the base sphere and the flow of the material of the mantle originated the initial fractures and the separation of the continents that now shape the world. When was this? De Beer, himself in his interesting Atlas, gives us the opportunity to identify, approximately, that moment.

In the Map 13 of his 'Map of Evolution' De Beer states: "Lungfishes are primitive fish that possess some of the characteristics that enabled their relatives to get out of the water and transform themselves, by evolution, into terrestrial vertebrates. These characteristics include the position of external *and internal nasal windows and of a lung to breathe air, from which these fish make use, when the water of the rivers in which they live dries. At present there are three groups of lungfish: 1 The Neoceratodus in the Mary and Burnet rivers of Queensland, Australia; 2 Protopterus on the White Nile, some of the great lakes, Oguburi, Congo, Zambese, Niger, and rivers of *Gambia, in Africa; 3 Lepidosiren in the Amazon Basin and in the Panama River of South America. The distribution of its fossils, characteristics of the Devonian period onwards, shows that the lungfish were originally universal and that the current restriction to their habitats is a consequence of their extinction due to competition in more easily accessible regions of the main center of evolution of the vertebrates in the tropical regions of the old world."*

When analyzing the aforementioned De Beer's map 13, it is clear that dipteridae's fossil have been found in Paleozoic geologic strata of Europe, Asia, North America, Australia and the Amazon. For this reason we coincide with De Beer that, in the Paleozoic the distribution of 'Dipteridae' lungfish was universal throughout the Globe.

In the Mesozoic, the Ceratodontidae fossil distribution is also universal; but, due to the effects of the extinction, at present, they are alive only in Australia. This suggests that the Ceratodontidae are an evolutionary stage subsequent to the Paleozoic Dipteridae and have been kept alive in Australia, where other forms as primitive as they also achieved the appropriate survival environment.

In the Cenozoic, all previous remains of Ceratodontidae (except Australia) have disappeared and the Protopteridae appear in Africa where today, its fossil remains have been found, but also living fish. Today in South America there are only Lepidosirenidae, both in their fossil form and alive.

It is therefore the transition from Mesozoic to Cenozoic, the geologic moment to which America and Euro-Africa continents arrived united, and the Atlantic Ocean still was a narrow lake.

Such world would be unrecognizable for the 'inter-temporal' traveler who achieve to identify it because he would find it different: smaller and hotter, since it would still conserve the energy not dissipated from the Mesozoic to our days; slower rotation with longer days, and shorter years appropriate to the smaller dimension of its orbit with lesser aphelion. In this condition Earth would has higher surface temperature in its sunny face as it would receive more energy for a longer time radiated from a closer and younger Sun. At the contrary the night side would be colder than today or perhaps a bigger ocean layer could maintain, by convection, a better heat distribution.

Climate conditions allowed and favored delicate life forms as the great reptiles whose existence is impossible in cold, dry and of great variability climates.

GRAPH # 6
LUNG FISHES

In Paleozoic, Dipteridae fish were of general distribution throughout the globe. Judging from the fossils found, this species was replaced by ceratodontidae. In Cenozoic all remaining Ceratodontidae have disappeared, except in Australia where fossils and live fish are found. In Africa Ceratodontidae were replaced by the Protopteridae, where today fossil remains have been found and also live fish; not so in South America where today the Lepidosirenidae are found, both fossil form and as live fish.

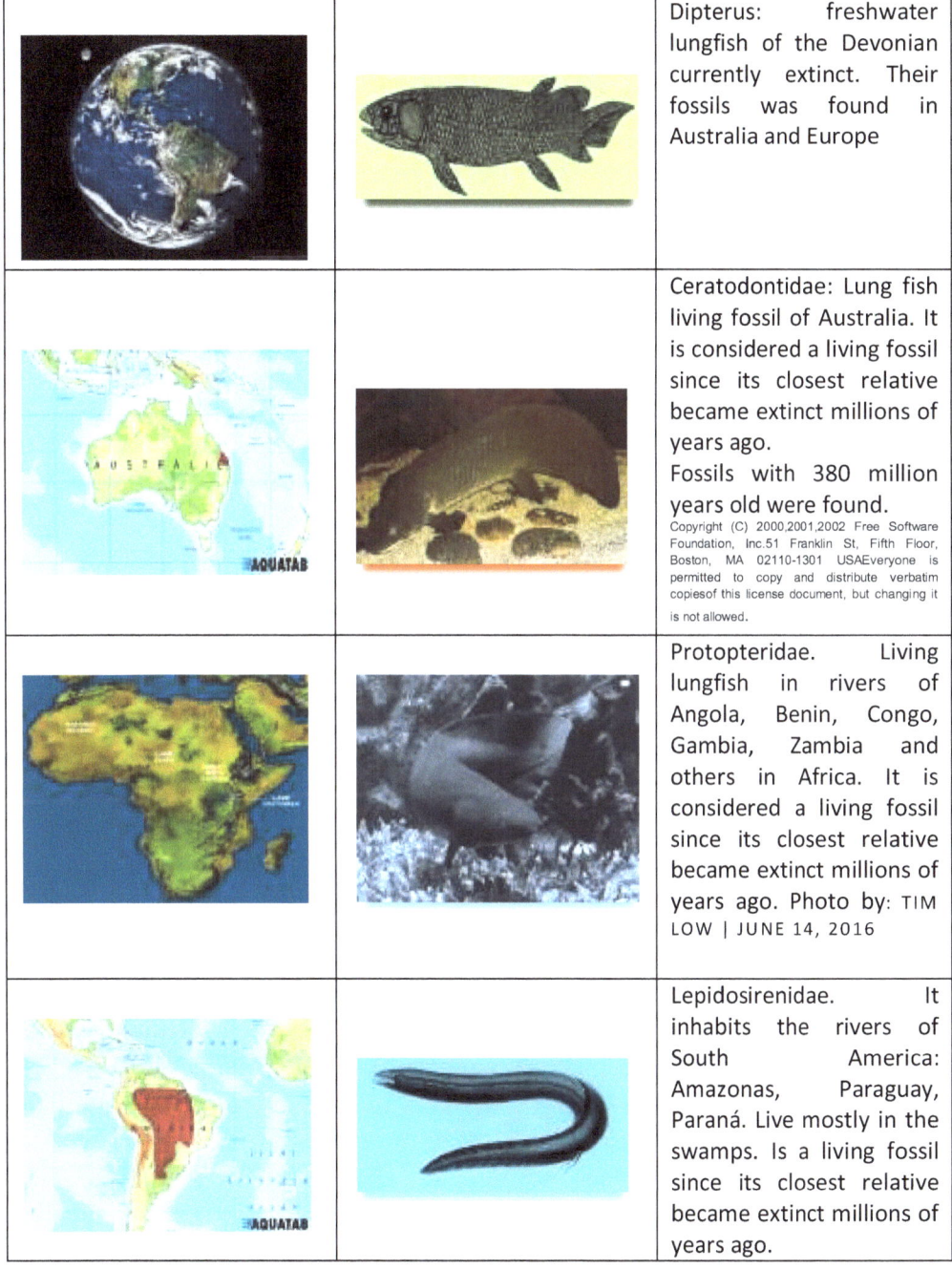

The proven existence of giant reptiles, such as Diplodocus or any of its congeners, forces us to accept that the Earth was a basically different planet in which, at the same time, plant species compatible with the dinosaurs that used them as food were developed. Those plants appear in the Devonian, the most in the Carboniferous and some in the Permian, Triassic and Jurassic, coinciding with the appearance of amphibians and reptiles, the same ones that ruled the Earth for several million years.

However, everything changed in a short period of time, today it is recognized that change happened during the transition from the Mesozoic to the Cenozoic. In other words, the planet Earth changed radically and all living beings, animal or vegetable, faced the transformation of their habitat and they adapted by overcoming the change or died. During that stage of radical changes more adaptable species developed. Those new living beings succeeded reptiles in the domain of the Earth.

The theory of the DYNAMIC BALANCE implies that all variation in the center of the system is reflected in the whole system. In other words, the planets will be affected in: orbital velocity, rotation speed, declination, composition of atmospheric and oceanic elements, planetary thermodynamics and many more. As it is logic, phenomenological interdependence implies variations in TIME. Consequently, we affirm that, coinciding with the planetary habitat changes, the "time" of living beings also changed.

In other words it seems possible that in the remote past, the same lapse corresponded to greater apparent times, compared to the current ones. This tells us that our days are shorter than the days of a dinosaur and that the Paleozoic world was closer to the young Sun and its heat.

7) THE CONTINENTS AND THE DYNAMIC BALANCE

The continents are original crust pieces that, driven by the constant spherical-expansive process, increase their radial distance to planet center and occupy a peripheral space in the expanded sphere. The continents are pushed to that position by the system of forces developed during the expansion; the system has two main components opposed to each other: 1. Spherical Internal Radial Pressure, produced by growth of the base sphere. 2. Structural strength of Bark pieces, which opposes expansion.

However, it must be added that current bark pieces shape and structure results from a long three-dimensional deformation process to which concur many other factors such as:

- The Radial Pressure developed by the Earth in its expansive process.
- The Earth crust structural resistance against expansion. This is a materials resistance phenomenon against deformation.
- Surface tensions on bark pieces that are developed in each expansion process. 'All force generates another force of equal magnitude and inverse direction'.
- The curvature difference between the original sphere and that of the crust, which they had the same curvature at the origin, but difference became greater with each

expansion. The consequence of this phenomenon is that the bark (rigid and inelastic) is constantly subjected to breach stresses in its central parts, subduction at West and traction to East. (See Graphic 9 Magma Tectonic Inertial Runoff)

- The crust pieces of greater or lesser dimension that resulted from the fracture. For example: a piece of lithosphere such as Eurasia will react differently to Greenland when it will be projected during expansion.
- The northern or southern geographical position of crust pieces, in relation with the transitory Equator. For example: the American continent was linked by its extremes to North and South poles, but an important part of current South America mass was and is mounted over the terrestrial Ecuador; it is logical that its response to expansion has been different from that of Australia that was always south of the equator and, in recent geological times, isolated from any other piece of the original crust, hence its structural deformation is lesser.
- The direction or the geographic angle of the forces that gave rise to the mountain chains. Example: In Eurasia, the East-West tension that gave rise to the mountain complex that goes from Spain to China; or in America the North-South tension that gave originated the complex of mountain chains that go from Alaska to Patagonia.
- The Earth rotational inertia. Example: The island arcs of the Philippines, Japan, Guam, South Sanduish Islands and others were formed in expansive processes influenced by the "Tectonic Inertial Runoff" that occurs when the rotation of the planet accelerates; but, the friction of the crust against the mantle slows down the cortical acceleration, this causes some magma to come out below the edge of the crust. That magma flue forms the insular arcs. Another effect of the planet rotation is the oceanic trenches that are also formed in expansive processes influenced by the "Tectonic Inertial Runoff".
- Erosion. Another important factor in the continents development is the erosion that is produced by wind and hydraulic flux. The product of this wear accumulates in sedimentary layers that, over time, reach enough dimensions to modify coastal profiles, watersheds, oceanic pits, land or marine chains and other accidents.
- Others not identified.

(See Plates 7, 11, 12 and 13)

At this point it is necessary to note that the forces that gave rise to the planet mountain ranges were, and still are, the response of the crust to the spherical growth of planet which, I repeat, expands in Dynamic Balance in consequence to losses of mass and energy, suffered by the planet system's nucleus and the system planets, as all EMEs in the Universe.

But, given that the crust had lost its structural continuity, the pieces or Paleozoic-continents underwent morphological modifications induced by factors such as: geographical position, terrestrial rotation, over-runoff, Tectonic Inertial Runoff, direction of tensions, and others.

In other words, the base sphere expansion forced runoff of the pieces of the original crust; is the case Africa and South America that have been separated from each other in East-West direction and have migrated towards the North; while the primitive Africa-Eurasia complex had been separated from North America and migrated to South.

It is important to note that, when the original lithosphere was divided giving birth to the oceans, the largest crust pieces were concentrated in the Northern Hemisphere; This fact caused that those remaining down south of Ecuador were cuneiform and had less resistant to slipping. The wedge shape is characteristic of Africa that separated from Antarctica, India that was "pulled" by Eurasia from the vicinity of Madagascar to its current position and South America that, initially linked to Antarctica, had a long process of subsidence that ended when its connection with the Antarctic Peninsula was broken.

8) Magma-Tectonic Inertial Runoff

During its expansion process, planets are subjected to all kinds of forces; one of them is generated by the rotational inertia that affects all the planet structures

In relation to crust and its parts (continents and others) it should be considered that, with any change of planetary rotational speed, the friction between Mantle and Crust will generate a force opposite to the rotational inertia, (See Graphic # 1 "Earth Expansion Process)

Magma-Tectonic Inertial Runoff is the process in which they combine: Surface tectonic cortical slip associated with Sub-flux of magma from the Earth's mantle. This process is generated by the moment of inertia that produces planetary rotational speed changes associated with changes in orbital speed. (Inertia Moment: is the body resistance to be accelerated in rotation).

As already said, the Earth rotation produces slippery effects that affect the oceans waters and atmosphere gases, the most known and studied consequences are tides and Carioles effect. However, the slippage effect also occurs between Mantle and Bark, but friction minimizes or prevents the "Inertial Magma-Tectonic Runoff" (See Graphics # 1 Earth Expansion Process and # 8 - Gulf of Mexico, Caribbean Sea and northern South America).

The runoff between crust and mantle is present during earth rotational acceleration phases, coinciding with expansion stages and as a result of an inertial difference that exceeds the friction between mantle and bark. As we saw in Book One; a higher rotating acceleration could cause the planet to "jump" into more distant orbit from system center.

GRAPH # 7
EARTH EXPANSION PROCESS
AUSTRALIA, PAPUA NEW GUINEA, NEW ZEALAND

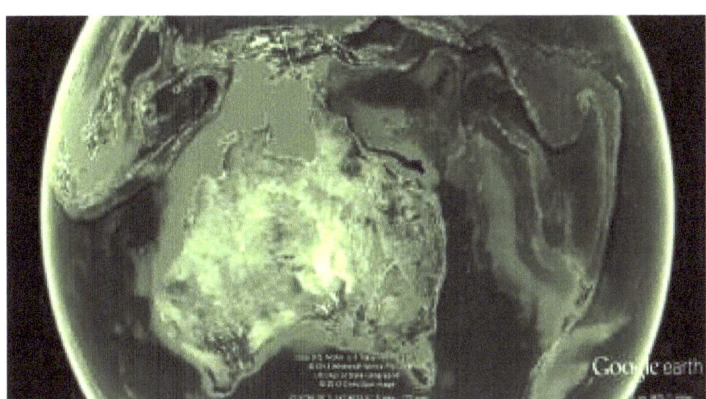

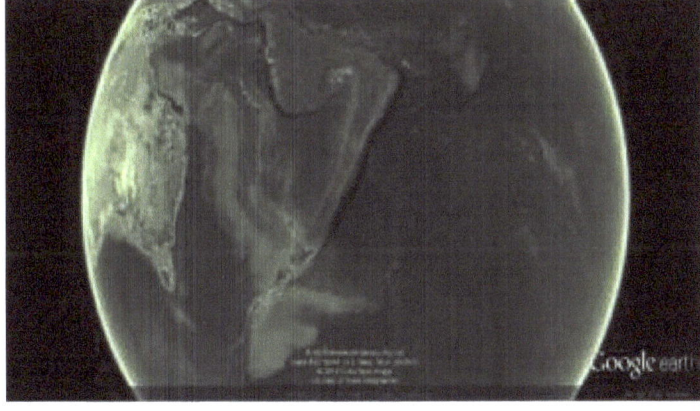

The GOOGLE view shows the structure of Australia and the set of geo-tectonic elements that surround it. The North Australian platform is separated by a long and narrow marine grave that encircles the Indonesian archipelago until colliding with the Northwest part of Papua New Guinea.

The described pit is continuation of the long perimeter that continues northwestward to the Andaman and Nicobar islands, north of Sumatra and southern coast of Myanmar (Burma).

Another element is that smaller island arcs, such as those that separate the Ceram Sea from the Banda Sea, have the same orientation as the arches of Japan and others; this suggests that its formation process was similar.

It is interesting to verify, in the photo of the seabed, that New Zealand is the southern part of a large structure that includes islands such as Vanuatu, Fiji and Tonga in the North; Adams, Auckland and Disappointment in the South, and others

In other words, as a consequence of the 'expansion'; the rigid and inelastic crust undergoes Tectonic Inertial Slip over the mantle; at the same time and under the rigid crust, a wave of magma arises that forms insular arcs at the eastern edge of the cortical piece.

Coincidentally, between the continental coast and the insular arc, an underwater plain extends that we will later call the sea; like the seas of Japan, of China or others. Examples of island arcs: Guam, Philippines, Japan, New Zealand, Caribbean Islands, South Sandwich

Islands (See Graphics 20, 21, 22, 23, 24, 25, 27 and section CENTRIFUGAL FORCE vs. CENTRIPETAL FORCE).

However, we must distinguish the Aleutian archipelago whose formation was due to the runoff of Alaska towards the Southeast and the Kamchatka peninsula towards the Southwest; while Eurasia separated from North America during opening of Pacific and Arctic Oceans.

**GRAPH # 8
LAND EXPANSION PROCESS
MEXICAN GULF, CARIBBEAN SEA AND PART OF SOUTH AMERICA**

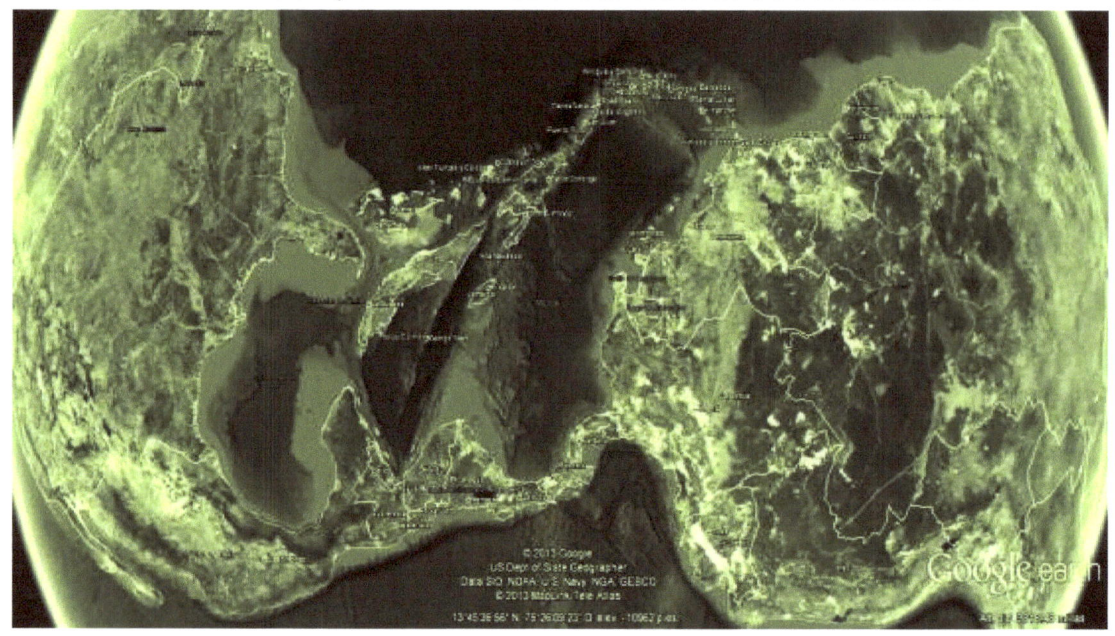

9) BELTS OR MAGNETIC BANDS

The lithosphere and asthenosphere vibrations, caused by the expansion, leave a MAGNETIC BANDS or BELTS trace that run parallel to the oceanic ridges. From the point of view of "Dynamic Equilibrium", those magnetic Bands are explained as follows. (See Graphic 10 - MAGNETIC BELTS)

The Earth globe expansion produces tension and compression stages of the Earth crust. At same time and along the oceanic chain, emergent material cools down in contact with the ocean floor water and acquiring "remaining magnetism", following principle:

GRAPHIC # 9
EARTH EXPANSION PROCESS
MAGMA -TECTONIC INERTIAL RONOFF

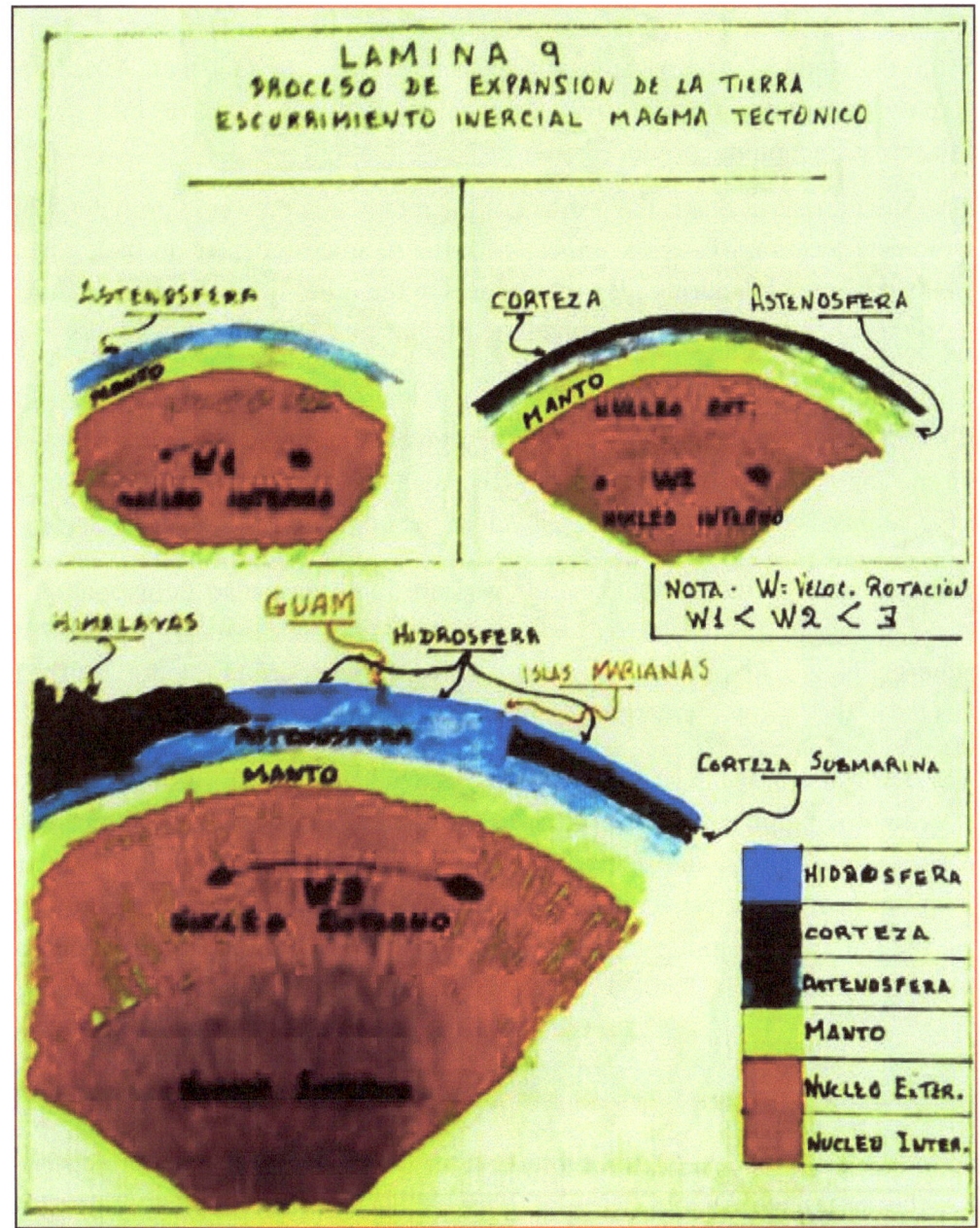

Any material immersed in a magnetic field, such as terrestrial, that has components capable of being magnetized and, at first, it is in a liquid or plastic state while subjected to vibration, will acquire residual magnetism when it changes its state, liquid or colloidal, to solid. Its residual magnetic polarity will be in one direction when the material vibrates in "compression" and inverse if it does in "tension".

This is more logical alternative to the "Planet Earth Magnetic Field inversion" that was widely widespread and accepted some time ago.

10) CONTINENTS AND EXPANSION

a) Africa

From the study of Africa geology it can be deduced the fact that, African plate has suffered fewer alterations than any other tectonic plate. On this continent would been initiated the crust fractioning, produced after its consolidation.

The Mediterranean Sea is the product of the interaction between: growth of the sphere and geographical location of Africa, which, with the geometric center on the globe equator, has maintained its position during all expansions. On the other hand, cuneiform Europe must flee to the northeast. In the middle remained the Mediterranean Sea in the process of growth.

The aforementioned pair of opposing forces, that gives origin the Mediterranean Sea opening; interacting with the alpine folding, they gives origin also the repeated open tearing of the Europe coasts starting with the Bay of Biscay, Seas of North, Baltic, Barents, Norway, and also the peninsulas Italic and Greek, as well as the opening of Black and Caspian seas.

Today, thanks to the "Google Earth" version 5 and its ocean floor photographs, we can see the run over trail, to Northeast, left by South Africa when separated from Antarctic, as consequence the growth process of the earth globe. It is important to note that the material that makes the seabed grow is emerging from inside the earth.

The African continent was radials projected when the terrestrial sphere expanded; given its geometric equilibrium position on terrestrial Ecuador, Africa was unable to slide towards the South or North and therefore has retained a geometric position similar to that occupied in the original terrestrial sphere. (It is important to note that the footprint left by Africa, on the seafloor, is similar to that left by India, at the bottom of Indian Ocean, when slipped to the North being "pulled" by Asia during the folding that gave rise to the Himalayas).

b) Asia and Europe

After the first expansion and rupture of the original crust of the terrestrial globe, Eurasia is the largest continuous mass of crust. The original Eurasia geometrical center position, led the Super Continent to move away from the Equator, towards North, with each new expansion.

However, the size of the base sphere grew so much that Eurasia, unable to cover the growing space created by the expansion, also moved away from the North Pole giving rise to the Arctic Ocean and the Great Atlantic Ridge. The Great Atlantic Dorsal is mountain ridge in spiral born near the North Pole and it extends towards the South, through a long curve whose most notorious point is what we know today as Iceland, from that point the direction becomes clearly south and marks the line of separation between Euro-Africa and America.

GRAPH 10
EARTH EXPANSION PROCESS
MAGNETIC STRIPS

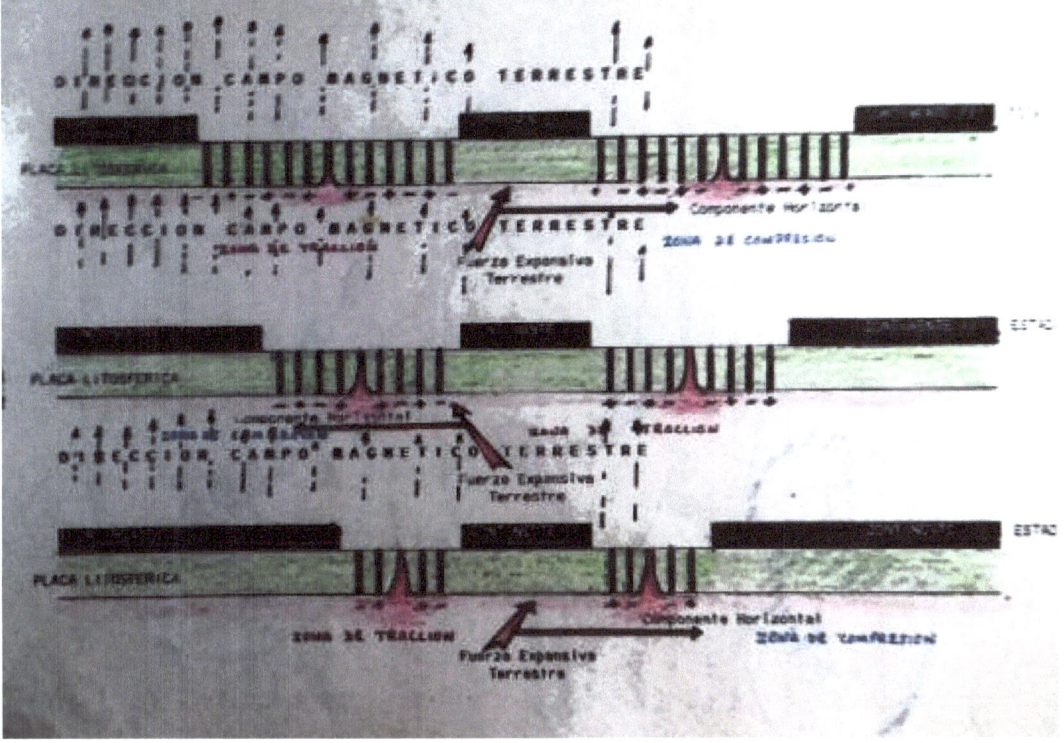

The enormous 'traction' (east-west and north-south) suffered by the continental plate Eurasia gave rise, as we saw before, to the mountain ranges that cross Eurasia in all direction and, in addition, to the folds that generated the rivers, Indo, Ganges and Brahmaputra whose controversial geological features have supported the assumption that India was an immense island adrift that traveled through the ocean of magma and reached its current position (See Graphic # 11 CORDILLERAS FROM SPAIN TO CHINA).

India with its wedge shape and geometric center north-equatorial offered little resistance and slid over the asthenosphere toward north, leaving the trace of its escape in the nascent Indian Ocean floor.

On the other hand, in the heart of Asia, the forces that gave structure and lifted the great plateau of Tibet and the chains of the Pamir, Hindu Kush, Karakorum, also opened the Persian Gulf, the Red Sea, and contributed to open the Mediterranean.

In addition, the Himalayan cortical folding exerted traction on: the Indochina Peninsula, the Indonesian archipelago, Manila and through them to Australia. However, the size, shape and geographical position of Australia, south of Ecuador, prevented its runoff in the direction of traction and all that large piece of bark separated and remained south of the Equator, as we see it today.

The accelerated expansion of the sphere forced the Euro-Asian plate to run to the North but I repeat, the huge plates of Africa and Australia were unable to overcome the equatorial limit and they offered resistance to also run to the North. However, due to the position of its geometric centers, the continuity of the Eurasian plate and their interconnection with Africa, break west side of continental Eurasian complex, mean while Australia did it from east.

c) America

In the Paleozoic, America was a continuous continental piece from the North Pole to the South Pole. This continental plate had the approximate shape of a large triangle, the first of whose vertices joined the South Pole through the present Antarctic Peninsula; the second vertex, very close to the North Pole, was strongly linked to the great Eurasian continental complex through Paleozoic -Alaska; the third joined to Eurasian complex by the European side, through Paleozoic-Greenland.

At that time, the Paleozoic-American continental plate had the same curvature as the planetary sphere to which it was still perfectly adapted. That is, the curvatures of the base sphere and the continental plate were equal. Under these conditions, the outer surface of Paleozoic-America, which ran from the South Pole to the North Pole, had had only altered by emergent elements such as volcanoes, there were no mountain ranges.

In other words, in the Paleozoic the American continental plate was a piece of the lithosphere with the same curvature as the colloid asthenosphere, on which it rested, and both resisted the increasing isostatic radial pressure, exerted by the expanding sphere. The result of this contrast of forces was foreseeable (See: Graphic # 12 Mountain Chains from ALASKA to PATAGONIA)

- The primitive crust's American continent, that had fixed points that prevented slipping, resisted the effort but stretched in North-South direction
- Additionally, it crumpled transversally in continuous and parallel wrinkles in direction of tension, as a piece of fabric does when it is stretched from its ends. That was the mountain ranges birth.
- The eastern end of the triangle it was separated from what would be Europe, Paleozoic-Greenland was born and the incipient Barry archipelago. All this while the north continental border slid towards south giving birth to what would be the Arctic Ocean.
- At south, the expanding sphere pushed the Antarctic continent still linked with America; this produced the stretched of Paleozoic South America in a north-south direction and the contraction east-west that gave birth to the Andes and the sinking of the South American southern tip. Later the union would be broken and the mountain range would emerge in a long progeny process that continues even by today.

GRAPH 11
EARTH EXPANSION PROCESS
MOUNTAIN CHAINS FROM SPAIN TO CHINA

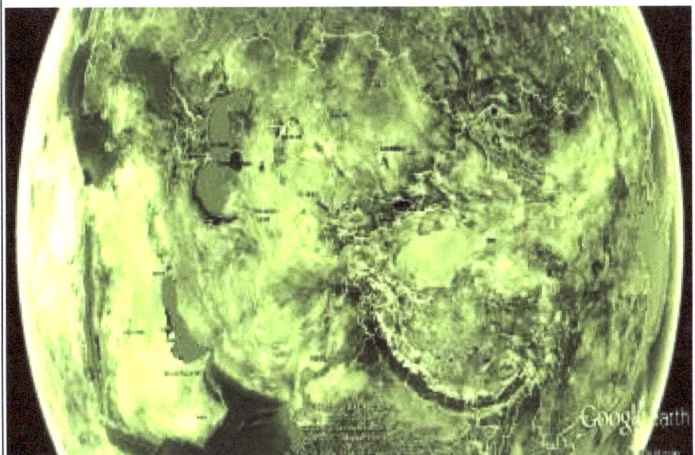

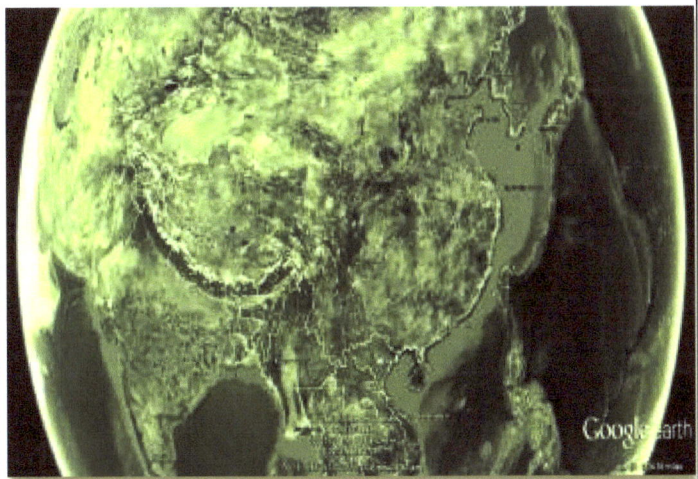

The great mass of China-Siberia-Indochina crust is contracted in a core of mountain ranges that attracts the continental edges. But, that continental piece is affected by the terrestrial rotation; for that reason the deformations are different in their western and eastern extremes. The West cracks in three peripheral seas: Mediterranean, Red and Persian. South is contracted and drag India to the North

The primitive crust deformation forms the Antarctic, Indian and Atlantic oceans embryos with their dorsal. The Pacific was already growing.

To the East it is contracted and Indochina is dragged towards the Northwest, it deforms and cracks in peninsulas and narrow islands. On China and Siberia side, kidney-shaped structures are formed under which magma flows forming island arcs as Japan archipelago and other.

The photo also suggests that applied forces composition over Arabia from Africa pulling South, the alpine fold pulling towards Northeast, and the North of Eurasia 'pulled' towards the Northwest, forced Arabia to turn West over an hypothetical center located in current Sinai Desert.

With this turn were opened: the Red Sea, the gulfs Aden, Oman and Persian, and made the Mediterranean grow, and the inland seas Black and Caspian.

It is important consider the incipient "dorsal" opened in the Gulf of Aden, in which cross-sectional fractures can be seen; that fractures are witnesses of lateral displacements.

- In North America case, the northern edge was broken with the following concomitant phenomena: the Hudson Bay was born, the Great Lakes were born, the crust was split giving rise to the Barry archipelago, Baffin Land and others.
- Most of the South American crust mass was, as today, mounted on the Equatorial Line. This condition forced that part of the continent to maintain its position but, given that the expansion process continued, South America distanced itself from North America by creating the Gulf of Mexico and the Caribbean islands. Central America was unfolded in a capricious manner and gave birth to the Yucatan Peninsula. When that Peninsula was ripped create the Gulf of Honduras. Panama stretched and unfolded, producing the valleys of Magdalena and Cauca in today's Colombia and Maracaibo Lake in Venezuelan land.
- It is very difficult to follow the complex deformation of the South American cortical piece in its constant adaptation to the changing curvature of the terrestrial sphere through the long process of expansion; but it is clear that, a powerful combination of very strong forces was applied in the area of the (today) Bolivian-Peruvian border; this phenomenon originated the violent deviation of the southern end of the Continent towards the West.
- This tectonic process also opened the basin of the Río de la Plata with their giants Paraná and Uruguay, as well as the Orinoco basin, the immense Amazon basin with its main Marañón, Ucayali, Putumayo, Caquetá, Negro, Tocantins, Purús, Madeira, Japurá, Xingú and Tapajós; besides having folded the Cordillera to form the highland plateau with the Titicaca Lake.
- At the America's south extreme, the interaction between the terrestrial expansion with the cohesion force of the crust and its mountain chains produced the definitive link rupture of Antarctic Peninsula and the Pacific edge of South America. (See Graphic # 21 PATAGONIA, ANTARCTIC PENINSULA, SOUTH SANDUICH ISLANDS).

d) Australia, Papua New Guinea, New Zealand, Indonesia

Australia, prevented from migrating to the North (due to its South-Equatorial position) but strongly linked with the northern cortical masses through the crust of what are now: Indochina Peninsula, the Malayan Archipelago, other islands and their tectonic platforms, it separated early from the Antarctic with which it was initially united. However, it remained south of the equatorial line.

Another connection consequence the between Continent Island with Asian archipelagos, was that, Australia remained in a constant meridian position, while New Zealand slipped to the East with its corresponding island shelf and other islands with which it formed one of the more primitive islands conglomerates of in the Earth. (See Graphic # 13 NEW ZEALAND AND SUBMARINE ENVIRONMENT) (See Graphic # 14 INDONESIAN PENINSULA, MALAYSIA, INDONESIAN ARCHIPELAGO, PHILIPPINES, PAPUA).

RAPH # 12
EARTH EXPANSION PROCESS
MOUNTAIN CHAINS FROM ALASKA TO PATAGONIA

NORTH AMERICA

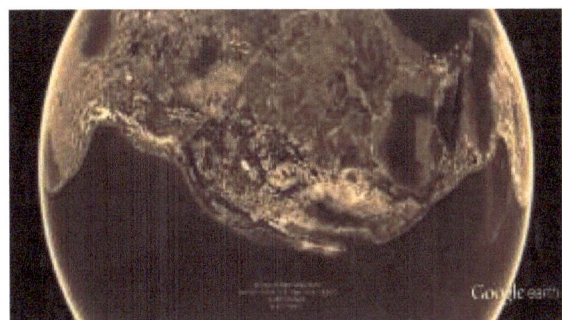

CENTRAL AMERICA

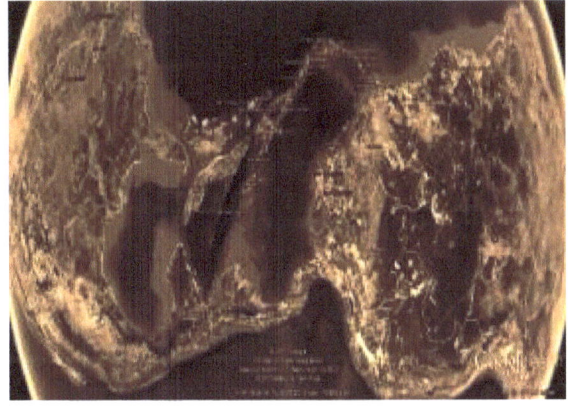

SOUTH AMERICA

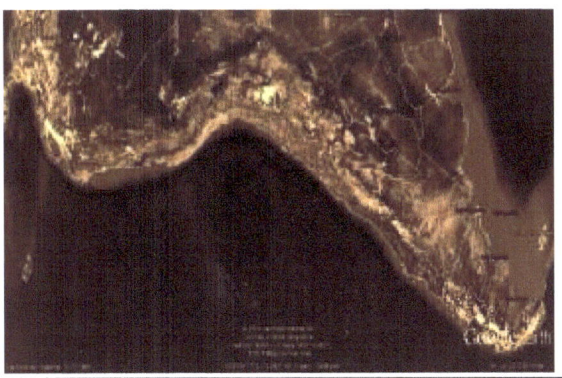

The impressive alignment of mountain ranges, from Alaska to Central America passing through Mexico is witness to the cyclopean cohesion forces that the crust developed against base sphere onslaught of growth. We can't ignore the Rockies area thickening, which is response to the transversal dimension of the continent.

Nor forget the meridian effort to maintain the structure, that didn't prevent the breakdown of Baja California peninsula, or discontinuity of Canadian coast.

The link between the two main continental masses is given by Central America and the discontinuous islands chain that go from Cuba to Venezuela; in middle, the Gulf of Mexico and Caribbean Sea. It is easy to imagine that the largest masses separated because one had its center of mass to the South of the Equatorial Line and the other to the North; this forced to stretching the link, unfolding all its structures.

The development of the Andean mountain range shows that the growth effort of the base sphere found resistance in the crust structure, this produced the mountain range. However, the direction of the traction pivoted the northern head of South America and produced the Bolivian Altiplano and Titi Caca Lake. On the continent opposite side, it stretched the coastline and opened characteristic gulfs and bays: Rio de la Plata, San Matías Gulf, San Jorge Gulf, Bahía Grande; in the Pacific side, the rosary of Chilean South islands.

It is interesting verify in the correspondent photo of the seabed that surround New Zealand, that this island is part of a large structure that includes islands such as Vanuatu, Fiji and Tonga to the North; Adams, Oakland and Disappointment to the South, and many others.

e) Antarctica

The constant tension produced by the globe expansion applied by millennia against the bark pieces cohesion force; produced the definitive rupture between the Antarctic Peninsula and the Pacific edge of Patagonia; that was the last link between Antarctic continent and any other continent. (See Graphic # 15. - PATAGONIA, ANTARCTIC PENINSULA, SANDUICH SOUTHERN ISLANDS).

As seen in the satellite photo, today the Antarctic continent has two well-defined masses, and its "Paleozoic-magnetic" pole is at the heart of the largest continental mass, displaced from its geographical equivalent by about 17 degrees.

The Antarctic Peninsula, almost turned off in South America direction, suggests that; during the process in which Africa and Australia crust's pieces was separated from Antarctica, the largest continental mass of South Pole was pulled slightly towards those continents; however Antarctica remained united to the American continent through the Antarctic Peninsula, which stretched like an elastic wire spring. Through millennia, this double displacement originated the drift of the Paleozoic-Geomagnetic South Pole to its current geographical position and helped, in America, to the long orogeny process that gave birth and growth to the long of mountain ranges that, as we have already said, goes from pole to pole.

11). Terrestrial Structures of Similar Tectonic Development

The geological structure of: Mexico Central America, Cuba, Caribbean islands, and northern part of South America, has an unequivocally similar tectonic development with the structure formed by: Indochina Peninsula, Malaysia, Indonesia, Philippines, Papua New Guinea and Northern Australia. (See Graphic # 16. - TERRESTRIAL STRUCTURES OF SIMILAR TECTONOUS DEVELOPMENT)

In other words, the global expansive force interacting against the cohesive force of the crust is expressed geologically, in both cases, in a similar cortical fractionation. It is clear that the difference between the cortical masses dimensions (Asia and Australia in one hand and North and South America on the other) as well as the orientation North South of America as opposed to the East West orientation of Eurasia, would explain the differences of dimension and others.

On the other hand, it is evident that the development of both structures has a Northwest-Southeast general orientation, due to the position of the large masses of bark to which they were connected in the past, and with which they still interconnect today. The two structures are the connecting bridge between large geological crustal structures. Some of seas or gulfs that are in there have zones of great depth, produced by fissures that cross their thin crust until almost touching the terrestrial Mantle.

GRAPH # 13
EARTH EXPANSION PROCESS
NEW ZEALAND AND ITS SUBMARINE SURROUNDING

By visually analyzing the underwater formations surrounding New Zealand, it is easy to see the similarity between them and those around Guam Island.

From this point of view it is possible to conclude that they are formations produced by Magma leaks due to processes of "tectonic inertial overflow" of great magnitude, during an Earth rotational speed acceleration event

What has been said confirms the expansion's theory approach, that is to say that these structures are part of the original crust and its development is product of the planet dynamic expansion.

From the above it is evident that the developments of these two areas have similar characteristics and when studying them must been taken into account factors such as: geological age, size and shape of the cortical structures from which they have been separated, transversal dimension prior to the stretching and others,

12). -AMERICAN AND EURO-ASIATIC'S CORDILLERA RANGE SYSTEMS

The mountain ranges are the terrestrial continental phenomena of greater significance and, according to the hypothesis that THE EARTH EXPANDS DYNAMIC AND BALANCEDLY, those structures contribute to explain how the original unitary crust was fractioned and today is a divided structure. According to the hypothesis, it must be kept in mind that the force generated by terrestrial sphere expansion is the origin and motor of the deformations and division. In turn, the expansion is consequence of the constant conversion of earth mass into energy that escapes to the universe. The planet internal mass transformed into energy produces reduction in gravity, this increases the volume of the base sphere that presses against the rigid crust. The gigantic surface tension, consequence of the increase in internal pressure through the ages, was reflected in lines of force that appeared on the surface of the globe, converted into mountain ranges. (See Graphics # 11 and 12).

GRAPH 14
LAND EXPANSION PROCESS
PENINSULA INDONESIA, MALAYSIA, INDONESIA, PHILIPPINES, PAPUA

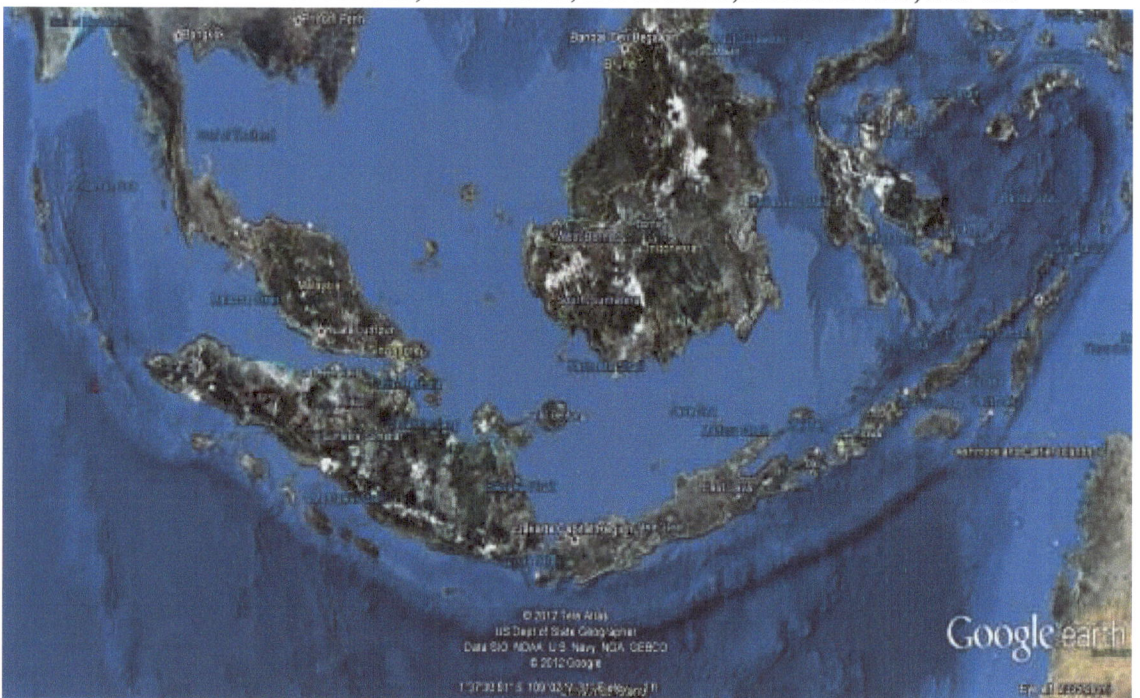

Considering: the terrestrial rotation, the inertial effect induced by the terrestrial sphere rotation, the terrestrial axis whose two ends define the current position of the geographic poles and the equatorial circumference that divides the sphere into two hemispheres; it is logical to accept that the tensions of the primitive crust, produced by the expansion, were mainly concentrated in two forces lines that they surrounded the globe: one from pole to pole in direction North - South, and the other in East - West direction halfway between the poles.

As it was predictable; as the sphere became larger, these tensions gave rise to folds that grew and cracked forming cordilleran systems. Those mountain chains systems were as larger and complex as larger were the pieces of bark in which they developed.

GRAPH # 15
EARTH EXPANSION PROCESS
PATAGONIA, ANTARCTIC PENINSULA, SOUTH SANDUICH ISLANDS

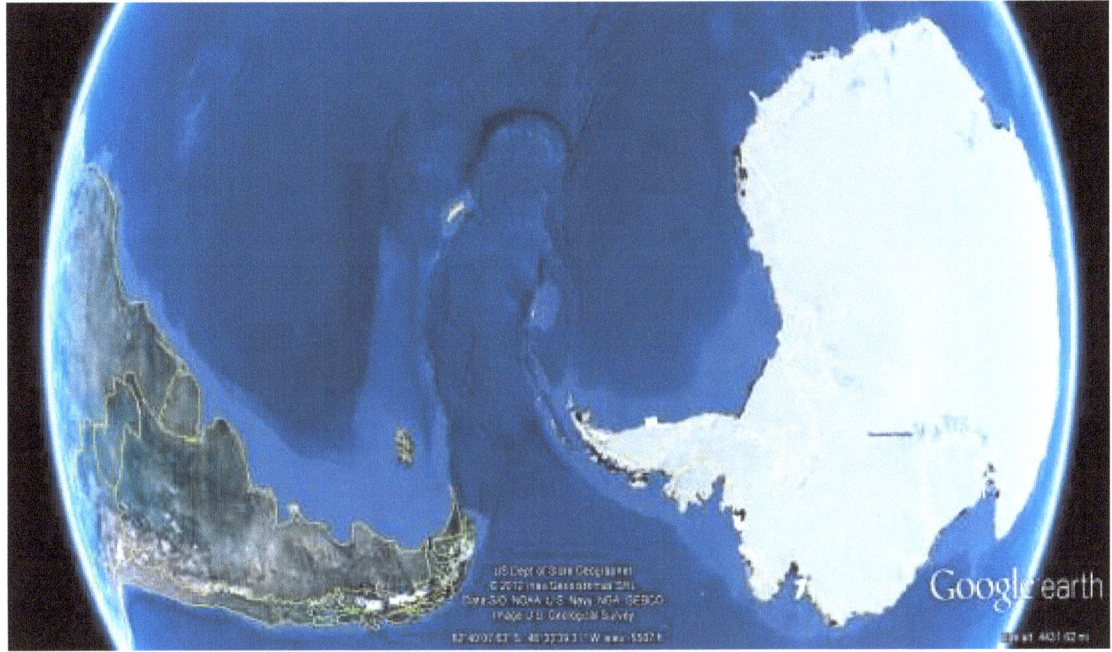

Given the "non-uniform" nature of the crust, it is legitimate to imagine that the mineralogical composition occurred in some more or less discontinuous distribution, this would explain some phenomena of formation or deformation. Example: the evident large copper deposits in the Peruvian-Chilean-Bolivian Andes, suggests that the tenacity of this element would have favored the stretching of south of the continent, through the ages, and its resistance to rupture with Antarctica

As we saw, the terrestrial sphere growth effort was transmitted to the primitive continents in spheroid-radial form but, at a given moment and for non-definable reasons, the forces were concentrated in two lines, they were:

- The Alpine Equatorial folding, with east-west direction.
- The Andean folding, with North-South direction.
- These two lines of force that crossed each other, impulse by the great effort of radial growth developed by the base sphere; it originated the escape of Alaska in the South-West direction, and the Siberia's extreme in a South-East direction. Besides, this displaced the northern border of America and Eurasia from a position near latitude 90° north, to its current latitude between the 60° and 70° north.
- For the America mountainous system, which goes from the North Pole to the South Pole (composed by the Andes, the discontinuous mountain ranges of Central America, the Sierra Madre Mexicana, the Rocky Mountains of North America and

the Brooks Range in Alaska), the "crust remote areas", which prevented the slide and generated the gigantic tension forces that raised the mountains, were and are:

GRAPHIC # 16
EARTH EXPANSION PROCESS
TERRESTRIAL STRUCTURES OF SIMILAR TECTONOUS DEVELOPMENT

A – Gulf of Mexico and Caribian Sea

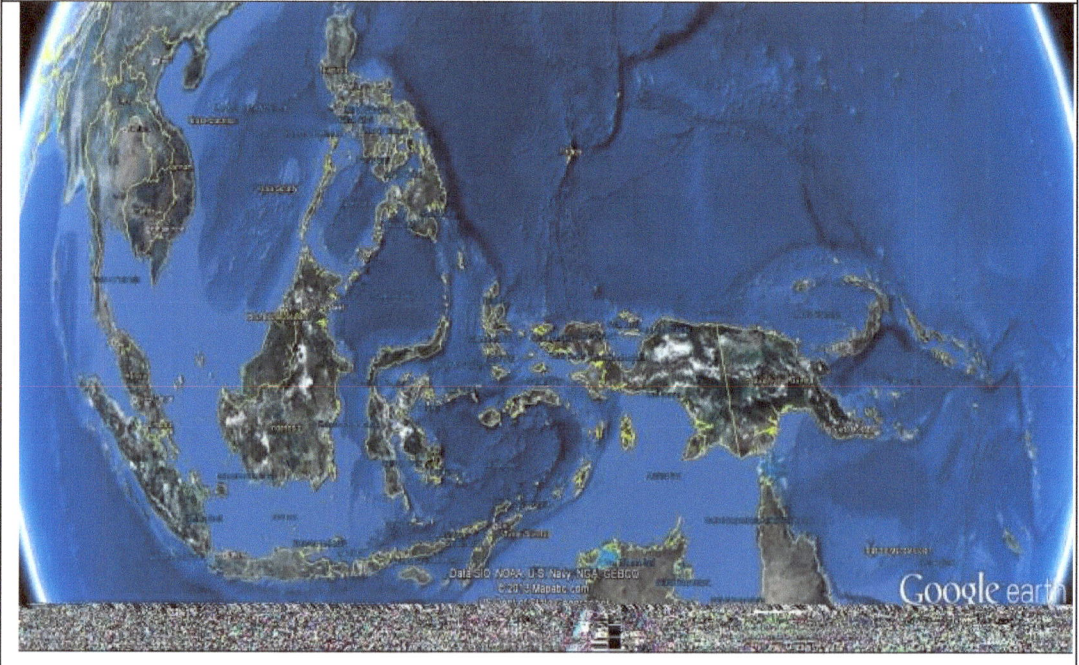

B - Indochina Peninsula, Malaysia, Indonesia, Philippines, Papua New Guinea and Northern Australia

- The crust piece of current North America, with its enormous dimension and the friction produced between its bottom surface and the corresponding asthenosphere zone, helped by its link with Asia by Alaska, and to Europe through Greenland, still united to the American continent.
- South America with two fixation zones; Patagonia strongly welded to Antarctica, and the great Amazonian surface on horseback over the Earth Equator.

Under these conditions American continent with its enormous spherical triangle shape, and subject to the global expansive process was deformed but maintained, during the expansion, an angle that goes from the meridian 178o West (in Alaska) to the 68o West meridian (in Patagonia).

The referred phenomena are the main factor in the continental cordilleran complexes formation, which, I repeat once again, are a response to the surface tension of the crust caused by the constant expansion of the terrestrial globe.

The global density decrease constantly, in direct ratio to the increase in volume and to the loss of mass and energy that escapes to outer space constantly. The decrease in density was reflected, and will be reflected in the gravity decrease and in successive expansions that will continue as long as the planet emits mass and energy, and continues to move away from its energy center (the Sun) as it was registered by the ancestral Calendars.

13). -THE OCEANS

a) The Oceans and their beds

With the advent of the space age and artificial satellites, arrived an unprecedented revolution in many areas of knowledge. One of those advancements, of great and spectacular impact, has been the spatial optics that allowed to photograph the oceanic bed; today we "see" it clearly in all their characteristics and accidents and, in addition, with the help of ultrasound devices "we know" with amazing precision the depth of each point of the ocean geography. Now it is possible review, compare and understand the theories of earth's crust geological dynamic (See Graphics #. - 15, 17, 18, 19A, 19B, 25, 26 and 27).

After the ocean floor panoramic analysis, through the GOOGLE EARTH photographs, it follows that there are accidents, different types of changes or modifications in the seabed, in this sense it is possible to identify, among many others, the following:

- Continental migration and magma spill areas, example: India, South Africa, Guam, Japan and the ocean floor of the Philippines

- Island arcs, examples: Sea of China, Sea of Japan, Kamchatka Peninsula, Indonesia, Bering archipelago, Solomon Islands, New Zealand, Caribbean Sea, Patagonia, South Sandwich Islands, and Antarctic Peninsula.
- Mountain chains or oceanic ridges, examples: North Atlantic dorsal and South Atlantic dorsal, Bismarck II fracture in the Indian Ocean, Pacific ridge.
- Oceanic rifts.
- Continental platforms

  Also, there are elements that affect those changes, modifications or accidents:
- Geographical position
- Nearby continental piece Dimension
- Connection or not with pieces of the original bark
- East-West or North-South development axis

On the other hand, the various types of changes, accidents or geological modifications it combines and complicate between them in many cases. It should be noted that the changes, accidents or modifications referred are also, in one way or another, a consequence of the terrestrial sphere expansion, which as we have seen, is the answer to universal gravitation when the EMEs loss its mass-energy.

The views of the ASIA-PACIFIC front presented in Graphic # 24 show that the island arcs formation is mainly due to a large amount of magma spilled out from under continental structures.

Those photographs from Google Earth show staggered spills of magma, which produced the island arcs of Guam and surrounding areas. This confirms that the island arcs are the product of terrestrial rotation acceleration caused the expansion events. These events are produced during variations of the inertial equilibrium.

The magma escaped below the continental crust was driven by inertia in an event in which the continent, firmly adhered to the base sphere, accelerated rotationally while the magma overflowing below the continental margin, leaving to our amazement the insular arcs that we see today, but that will be repeated in the future with each expansion process with acceleration.

In Japan archipelago, the underlying magma ripped and overflowed the continental limit in successive of Earth's expansion and revolving acceleration stages; this gives birth to insular arch and Japan Sea with its bed. (See. - Graphic # 23).

GRAPH # 17
EARTH EXPANSION PROCESS
PACIFIC OCEAN, BEDS NORTHERN, CENTER AND SOUTH

NORTH PACIFIC OCEAN

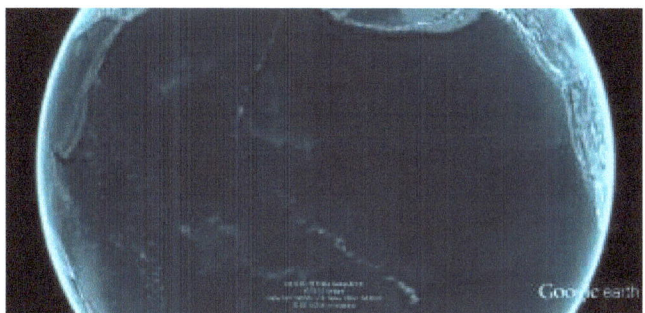

CENTRE PACIFIC OCEAN

SOUTH PACIFIC OCEAN

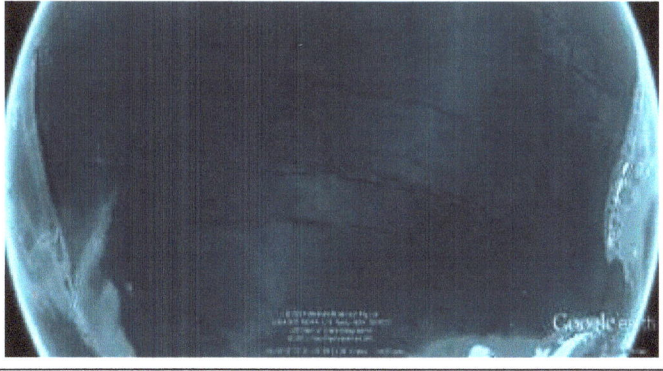

NORTH PACIFIC OCEAN

The GOOGLE image shows an over-runoff trace, which turns into volcanic protuberances. That protuberance is Hawaii Island; the trace moves northwest to a point at Latitude 32° 21 'N Longitude 173° 38 E, where it changes to the North-Northeast until it meets Kamchatka.

On the East side there are traces of North America slippage. The most notorious traces are three with orientation from West to East.

At west of the volcanic line there is a large area with discontinuous protuberances, obeying no pattern and suggest volcanism stages.

Johnston atoll is part of a parallel structure to the Hawaiian formation that includes Palmyra atoll and Kiribati island and reaches Tuamotu Islands "Ridge" in French Polynesia.

CENTRE PACIFIC OCEAN

From Tuamotu ridge there is up to seven lines of slip that extend toward South America, suggesting that the terrestrial sphere and the Pacific Ocean expanded as that continent slipped through millennia to the East.

SOUTH PACIFIC OCEAN

In this part the previous characteristics are repeated; on the line that extends from East to West at the height of the Tropic of Capricorn is French Polynesia; further east Pitcairn, Easter and, near the Chilean coast, San Felix and San Ambrosio. From that latitude to the Antarctic the last four major expansion lines are seen.

In the cases of Bismarck Sea and Indonesia, the case of migration and escape of magma with formation of island arcs is repeated, as shown in plates 20 and 21. In this case, however, it should be noted that the island arcs they formed with part of the expanded continental margin and, after the arches, the continental platforms form depressions that in future expansions could constitute interior seas, gulfs or bays. The views of the Bering archipelago with its sea and Patagonia with the Antarctic Peninsula and the ocean floor are a demonstration of two geological phenomena that occurred during earth expansion events in distant geographical areas; however, the events had similarities and differences that we can see today. (See Plate # 22)

The most outstanding difference is that, while South America dragged and forced the slide of Patagonia towards the Northwest, the Antarctic Peninsula and the ocean floor developed in a West-East direction; this generated that the rotating acceleration of the Earth impelled the magma towards the East; today we can see the South Sandwich Islands on that magmatic ridge.

On the other hand, the development of archipelago and Bering Sea has a general North-South direction, since, when the Atlantic Ocean was opened, the Arctic was also opened. The Paleozoic-continents America and Euro-Asia separated by migrating to the south and, as they moved, several important changes occurred. Alaska turned to the southwest, contributing to the opening of Hudson Bay. On the other hand, the Kamchatka Peninsula turned towards the southeast originating the today Okhotsk Sea. The stretching force that separated the continents originated the Bering insular chain that defines the southern edge of the continental shelf of the original terrestrial crust in that area.

b) Tonga Trench

On 3/6/2013 at 9:04 PM, I observed a point of great depth in the vicinity of Tonga Island. The observation was made with the "Data LDEO Columbia NSF NOAA" program through GOOGLE EARTH. The geographic coordinates were:
22° 22 '18.38 "S - 174° 10' 34.91". O

The origin of the ocean trenches obeys various geodynamic reasons.

The expansion of the terrestrial sphere is through successive stages that correspond to bigger and bigger terrestrial diameters, consequently, the curvature of base sphere is more and more open with respect to the primitive crust curvature that, due to its rigid nature, maintains its original curvature. (See Graphic # 1)

As a consequence, when increases the planet diameter, the bark pieces edges press on the asthenosphere with increasing force. The surface pressure exerted by the growing seabed is added to the previous phenomenon. This double effort accompanied by other phenomena, such as the "Magma-Tectonic Inertial Overflow" effect, gives rise to processes of subduction and ocean trench formation.

GRAPH # 18 EARTH EXPANSION PROCESS ATLANTIC OCEAN CENTRE, SOUTH AND BED	
ATLANTIC OCEAN CENTRE  ATLANTIC OCEAN SOUTH 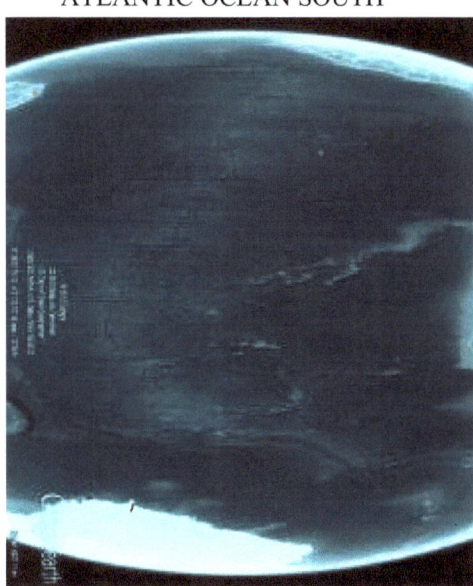	ATLANTIC OCEAN CENTRE The Atlantic's Dorsal runs southward in the central part of that ocean. It is to note that the continuity of the line, which was characteristic of the North part, is lost in this section. When arriving at Virgin Islands meridian, the dorsal line changes; it continues to the South adopting form of a broken line crossed by transversal fractures, every certain distance. The main line of the Dorsal remains halfway between Africa and America. The broken line jumps allow it to follow the continents contour. In this section is the Ascension Island, referred in section "A Chronological Attempt ATLANTIC OCEAN SOUTH The Atlantic Dorsal continues to South and ends in Lat. 50° 41 '86 "S, Long. 6° 37 '19.25 O; depth 8650 feet. It is important to note that in this area the Atlantic's bed is altered by elongated outcrops linked to the coast of South Africa, and others on Patagonia and the Antarctic Peninsula sides. In one of these outcrops there is a group of three seamounts identified as: Hints, Sandaled, Meteor (there are other)). Making use of the GUGLE EARTH, author has verified what appears to be another mountain, or the extension of one of the named, its location would be Lat. 47° 39 '27.86 "S, Long. 10° 22 '11.66 "E. The depth would be 9 feet. This mountain is surrounded by ocean depths of more than 14,728 feet.

Another type of trench arises when mountain ranges, bordering continental masses (like the Andes in South America) are stretched in response to global expansion. In this case, the transverse contraction at 90° of mountain range axis makes the mountains range grows vertically. This produces subduction and gives rise to the ocean trench.

GRAPH # 19 A EARTH EXPANSION PROCESS NORTH INDIC OCEAN AND ITS BED	GRAPH # 19 B EARTH EXPANSION PROCESS SOUTH INDIAN OCEAN
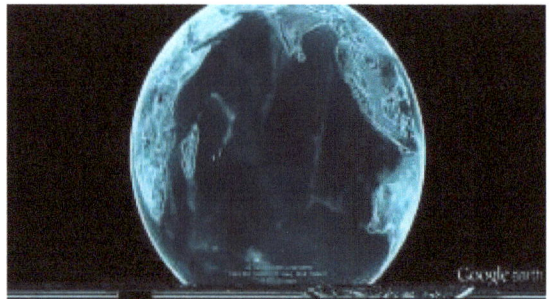	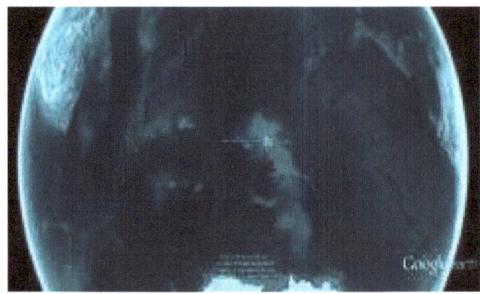
The dorsal that passes between South Africa and Antarctica and crosses part of the Indic Ocean to Northeast direction, varies to North towards the West coast of India and in this section are the Maldives islands. The submarine mountain range that goes from Indic Ocean's British Territories towards North and runs parallel to the western coast of India is another element that shows that India has escaped to the North dragged, from a position near the Ecuador to its current position, by the Eurasian continent to which was always belonged.	The GOOGLE photo shows the Indic Ocean Southern part, in it we can see two submarine ridges that open, one towards West to pass between Africa and Antarctic, and the other towards East to pass between Australia and Antarctic. At center of both are the so-called French Southern and Antarctic Lands and the Heard and McDonald Islands crowning a structure that appears to be, magma outcropping occurred when Antarctica was left behind while the continents Africa and Australia fled north during an event of terrestrial expansion, or is a small part of the original crust that was left behind and grew with volcanic magma.

c) Oceanic mountain ranges

The oceanic ridges are also result of expansion. As the terrestrial balloon expands, cracks appear in the thinner part of the ocean floor, generated by the tension produced by the expansion, the plastic magma at high temperature flows from the asthenosphere through the cracks. The accumulation of magma and its rapid cooling give had given and will give origin to the chains that form the oceanic mountain ranges. The most remarkable oceanic mountain ranges are in the Atlantic Ocean with a general North-South development; instead those of the Pacific and Indic have more complex but less defined development.

d) The Cocos Plate (See Graphic # 24 COCOS PLAQUE)

The Graphic # 24, shows a photography of COCOS Plaque sea bottom, around (GOOGLE EARTH geographic position Lat. 9° 55 '38.48" N - Long. 104° 31' 04.72' W), the white points that can be seen at west of the volcanoes are places of scientific exploration. The average depth around the study points is more than 8350 feet. The mouth of the largest volcano is at 500 feet depth under sea level, which gives that volcano 2850 feet height average above the ocean bed. The North-South lines near the volcanoes are a response to expansion events.

e) Pacific Ocean (See Graphic # 17 - PACIFIC OCEAN NORTH, CENTER, SOUT, AND THEIR BED)

The Pacific Ocean born as a the consequence of the first Earth expansion after the crust consolidation, the fracture of the crust ran from pole to pole and created the first oceanic pit to which part of the waters that covered the entire globe was precipitated. The Pacific Ocean bed is crossed by a complex chain of underwater mountains that are the product of diverse expansions which marked the bed of this ocean. \

GRAPH # 20 — EARTH EXPANSION PROCESS — INSULAR ARCHS, ANGULAR ACCELERATION'S PRODUCT AND SLIDING'S MAGMA UNDER CONTINENT'S CRUST IN EXPANSION EVENTS		
THE INSULAR ARCH OF THE GUAM ISLAND, IT IS NOT THE ONLY ONE IN THAT AREA AS SHOWS THE OTHER TWO PHOTOS OF THIS GRAPH. White dots are oceanic islands	THE PHOTO SHOWS MAGMA'S ESCAPE WHICH GAVE ORIGIN TO GUAM. IT IS THE OLDEST AND WAS FOLLOW IT, AT LEAST, BY TWO MORE. IT ALSO SUGGESTS THAT JAPANESE ARCHIPELAGO WAS PART OF THOSE GEOLOGICAL EVENTS. White dots are oceanic islands	IT IS INTERESTING THAT THE NEAR INDONESIA AREA PRESENTS SIMILAR SYMPTOMS. THIS SUGGESTS SAME CHARASTERISTICS THAN PREVIOUS TWO.

f) Atlantic Ocean (See Graphic # 18.- ATLANTIC OCEAN, CENTER, SOUTH, AND THEIR BED).

The Atlantic Ocean is second in dimension. It is crossed by a chain of underwater mountains that runs from pole to pole following a general north-south direction called "Atlantic Dorsal".

The Atlantic Dorsal mountain range begins in the Arctic Ocean at a point whose approximate geographical position is: Lat. 81° 38 '31.70 N, Long. 119° 17, 39.07 E, and 15559 feet depth. That point is at short distance from Volchevic Island.

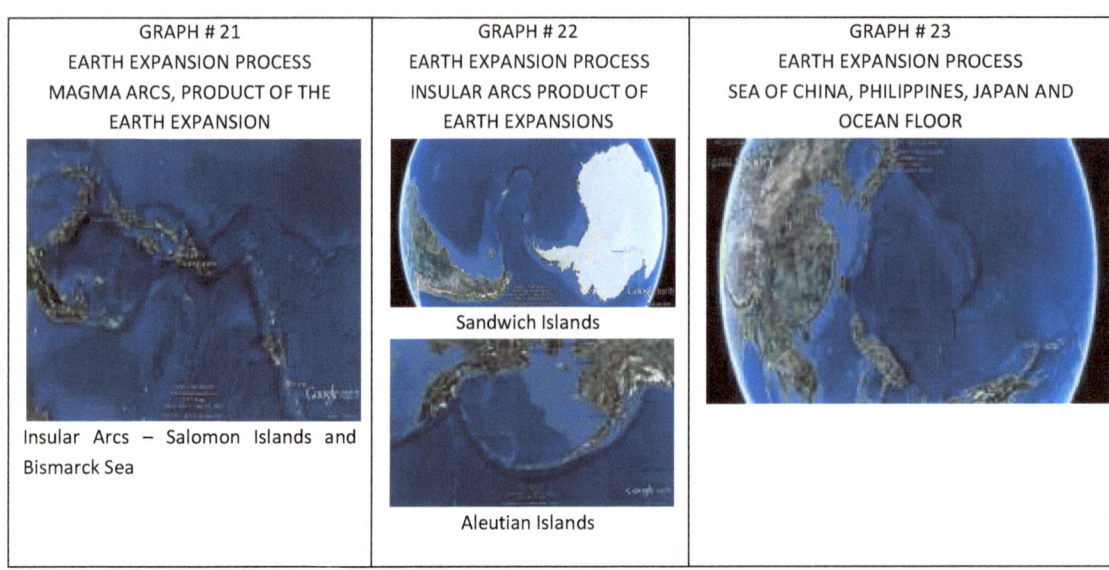

The dorsal one develops towards the South and passes between Greenland and the Norwegian islands Svalbard. Continue toward south until reach Iceland that is on the mountain range ridge; evidence that situation with its important volcano-thermal activity.

The Atlantic Dorsal ridge continues towards the South following the profile of the continents that limit the Atlantic. On the Dorsal there are characteristic islands that are in this work in section "A Chronological attempt".

g) Indian Ocean (See Graphic # 19A, 19B – NORTH and SOUTH INDIAN OCEAN AND its BED, Graphic # 25 - ARABIAN SEA)

The Indian Ocean is the third in dimension, and its North limit is the Asian continent. All its northern part is among the tropics. The Arabian Sea to the West of India and the Gulf of Bengal to the East they has its northern limit very close to 23° 27 'North (Tropic of Cancer), while the Tropic of Capricorn (parallel 23° 27' South) is to the South of Madagascar in the West and passes through the Cape Range (Australia) East of the Indian Ocean.

The southern part of the Indian Ocean reaches the Antarctic Continent and communicates by the East with the Pacific Ocean, at South of Australia; and by the West with the Atlantic Ocean, at South of Africa.

The Indian Ocean has its own characteristics, different from those of the other two great oceans. It shows with great clarity the interaction between the continents and the ocean floors, meanwhile that, the planetary sphere got bigger. It is absolutely clear that the Indian Ocean is circumscribed by four continental structures that moved away from each other throughout the millennia.

LAMINA 24
PROCESO DE EXPANSION DE LA TIERRA
PLACA COCOS

Uno de los extremos característicos del Pacífico central es el ángulo formado por el litoral de Panamá y Costa Rica con los de Colombia y Ecuador. En la fotografía del fondo oceánico se aprecian dos líneas de sobre-escurrimiento que, naciendo del archipiélago Galápagos llegan a las costas de los mencionados países. De acuerdo a la teoría de expansión del globo terrestre, es claro que los puntos donde llegan las líneas de sobre-escurrimiento se han alejado de las Galápagos en el remoto pasado, en tanto que Centro América se desdoblaba para mantenerse unida a las dos Américas, y América del Sur se alejaba radialmente hacia el Este de las islas, pero manteniendo su latitud sobre el Ecuador terrestre.

Las características de la placa de Cocos se pueden ver en su foto del lecho marino, las líneas en dirección norte sur muestran que la expansión, en esa zona, ocurre de oeste a este debido a que el continente corre hacia el este. La posición de América del Norte en relación con América del Sur obligó a América Central a extenderse en dirección NW hacia SE
La expansión, la aceleración de rotación terrestre y el afloramiento de magma fueron los motores de esos cambios

- To the North: Eurasia in the part corresponding to coasts of Arabia, Iran, Afghanistan, and India.
- To the South: the Antarctic continent, with its ocean through which the Indian Ocean communicates with the Pacific and Atlantic Oceans.
- To the East: Indochina, the Malay Archipelago islands and The Western Coast of Australia.
- To the West: the East coast of Africa

From the submarine structures analysis, can be seen mountain ranges produced by outcropping of magma with islands such as the Maldives, Seychelles, Mauritius and Reunion, French Southern and Antarctic Lands, Heard and McDonald Islands and others.

There are also several dorsal structures; some with stepped shape, others with unidirectional shape and lines of expansion parallel to the mountain range, all of which made it possible and in the future will facilitate the displacement of adjacent pieces of bark, during spherical-radial terrestrial expansion processes.

h) Antarctic Ocean (See Graphic # 26 - ANTARTIC OCEAN and Bed)

The Antarctic Ocean surrounds the Antarctic continent and communicates with the larger oceans, Atlantic, Pacific and Indian. Its bottom has mountain chains and dorsal that left the huge forces that moved the continents Africa, Australia and America far away of Antarctica, responding to the terrestrial globe expansion and its rotation speed changes.

By observing the structure of the Antarctic bed, the narrowing of the Drake Pass is obvious. The oceanic circulation which turns from west to east, forces the oceanic water to pass through the narrow channel that forms between Patagonia and the Antarctic Peninsula. This characteristic generates the strong and irregular currents that have made navigation difficult from the time of Colon discovery until our days, but also facilitate the energy transfer by convection through the oceans that is increased by circulation.

It is important to consider that the thermodynamics of the Earth's oceans varied when the expansion broke the link between Patagonia and Antarctic Peninsula. In times before that break, the ocean waters had fewer communication paths, had less mobility and ocean thermodynamics were different from today, could those geological changes explain radical climate changes, such as the ice ages of the past?

i. Arctic Ocean (See Graphic # 27 - Arctic Ocean and Bed)

The primitive crust around North Pole was fractured when Eurasia and North America were displaced to South during Earth expansions events. Those events had concomitant phenomena such as the fragmentation of the Barry archipelago, the incipient separation of Greenland, the opening of the Hudson Bay, the birth of the proto-Atlantic and, between Alaska and the Kamchatka Peninsula the Aleutian Island Arch. In the oceanic bottom photo, some important accidents can be identified.

The Atlantic Ocean dorsal was born near the North Pole extends from the Arctic to the Antarctic and, as it was said when speaking of the North coast of the continents Eurasia and America; if we compare the current position of the geographic North Pole (marked with green icon in the photograph), with the magnetic remnant of the Paleozoic-Polo, which is located in one of the islands of the Barry archipelago (marked with red icon in the photograph); it is congruent to accept that the magnetic remnant of the Paleozoic-North Pole, migrated dragged on the part of the continental crust of the t in which it was before the opening process of the Arctic Ocean. At the same time the coastlines of North America and Eurasia migrated to the South. However, the opening was not the same on the entire American-Eurasian coast, but it followed a "spiral pattern". Therefore, Alaska and Kamchatka is shelf connected by a common plinth; while at the other end of the Arctic, the Atlantic undersea mountain range it is at middle between Greenland and Europe.

GRAPH # 25
EARTH EXPANSION PROCESS
ARABIC SEA

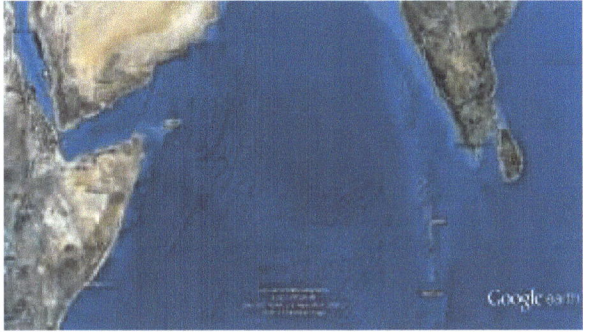

ARABIC SEA

The Arabian Sea is part of the Indic Ocean, located at India West side. See the Google photo, the South boundary is given by a dorsal that runs from West to Southeast, from the island Suqutra (Yemen), until the height of Maldives islands. Then it runs parallel to of Maldives islands dorsal. The final part of the ridge has staggered displacements similar to the mid-Atlantic ridge. This demonstrates the transversal effort of crust adaptation to the spherical Earth growth. To the Southwest of those displacements there is a submarine angular formation with the Seicheles islands to the North, Mauritius and Meeting Island at South. This structure suggests an ancient outcrop of magma. Some more to South is Madagascar

GRAPH # 26
EARTH EXPANSION PROCESS
ANTARTIC OCEAN AND BED

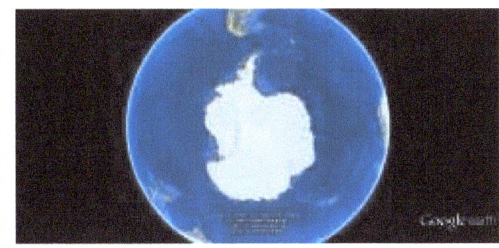

GRAPH # 27
EARTH EXPANSION PROCESS
ARTIC OCEAN AND BED

CONCLUSION

Having as a premise that the expansion of the globe has generated important changes in the geology of the continents and their cortical structures, it is evident that the Euro-Asian continent has moved away from the North Pole but, it has also distanced itself of the equator because the surface of the base sphere became larger than the size of the original crust, this was same in each original pieces of crust.

For this reason the EURO-ASIA FOLDING has served as a large belt, forcing that part of the bark suffer an important superficial narrowing. The mountain ranges of North India prevented the Indian Ocean to cut the continent for to continue its way to the North, so that was locked in between continental pieces of bark, as we know it today.

On the other hand, the TERRESTRIAL GLOBE EXPANSION, the ANDEAN FOLDING, and the ATLANTIC DORSAL produced the meridian division of the primitive crust and gave rise to the great Atlantic and Pacific oceans.

The phenomenon that this essay calls "Magma Tectonic Inertial Overflow" clearly explains the existence of Insular Arcs such as Guam and others, but also opens a research front against which many of the ideas that science has been accepting must be submitted, until today.

The "data" provided by the ancestral cultures calendars open new research spaces, I believe that its evaluation and study is essential.

Another important point is that, India was "dragged" to North when Eurasia had to stretch and adapt to the expanding terrestrial sphere. It is evident that the tracks left by the Sub-Continent in the Indian Ocean bed were produced by a large geo-tectonic piece that was "sliding by traction" and was not "pushed".

Having on mind all write, summarized in:

a) The constant decrease of the Solar 'Mass', which is reflected in the constant decrease of Gravitational Attraction force over the planets,
b) The tracks left by the pieces of crust, such as the Indian Sub-Continent, Africa and the different forms of mountain chains and submarine ridges on the oceans and seas beds,
c) The chains of terrestrial mountains in Eurasia, continental America and others,
d) The "Magma Tectonic Inertial Overflow", ignored by science until today,
e) The calendars of ancestral cultures,
f) The crust superficial contraction, produced as a global expansion consequence, that generates tension forces capable of moving large crust pieces; as the case Indic Subcontinent when it slipped on the earth's mantle dragged by the Eurasia longitudinal deformation;

Must be accepted; that the Earth "balanced expansion" was, is and will be real.

Consequently, the Earth will continue to expand through the ages if the Sun continues to lose mass; but, as we will see in "Book 3" of this work, human action has altered terrestrial thermodynamics and is changing the earth's crust dynamic equilibrium with serious risk for all living beings on OUR planet.

EARTH THERMODYNAMIC EQUILIBRIUM

THE EARTH, A THERMAL MACHINE

Author: LUIS JAVIER ARTIEDA CARPIO

BOOK THREE

EARTH THERMODYNAMIC EQUILIBRIUM

In Book 2 we saw that the Earth is a heavy nucleus planet, with an internal high energy level. The energy is transmitted to earth surface through a convection process that is difficult to study, define and evaluate but to which they converge: the terrestrial morphology and its discontinuities, variation of nature and phases, energy orientation, the changes that energy produces in its transit from the center of the sphere to the surface.

As it is congruent, the energy transiting inside earth affects: the Earth's Magnetic Field and the mass/energy that leaks through space, tides and their consequences, especially the Tectonic Seasonal Tide whose nature could alter the terrestrial balance with serious risk for living beings still living on the earth and many other effects until not studied.

On the other hand and discounting the huge seismic file we have very little information about the energy that generates earthquakes. Despite these limitations, geologists have come to the conclusion that planet is divided into several concentric parts called: inner core, outer core, mantle and crust. These different parts are separated from each other by what is now known as "DISCONTINUITYS".

1) DISCONTINUITIES. WHAT ARE THEY?

Through the study of terrestrial seismic, including data artificially produced by underground nuclear tests, it has become evident that there are certain limits separating the terrestrial sphere in layers that the human being has called DISCONTINUITIES.

These limits are enveloping and continuous geometrical locations, of the sphere, where thermal and / or physical changes of the earth material occur. These changes are evident due to speed alterations of the seismic waves that cross them.

There is no reason to suppose that the material of the discontinuities has different nature from the material that is before or after, but it is necessary to accept that they mark the limit where the terrestrial mater changes **phase or state**. When seismic waves pass through discontinuities, they modify their behavior consistently. This has led seismologists to build a form of structural radiography of the planet.

For the purpose of this work we will focus on the discontinuity closest to the earth's surface that has been named after its discoverer, ANDRIJA MOHOROVICIC.

(What is a phase? - A phase is a homogeneous part of a system, which, although in contact with other parts of the system, is separated by a well-defined boundary. It is a

material's region with homogeneous physical and chemical properties, a region that differs from another, in microstructure and / or composition.)

GRAPH # 1
BALANCE IN TERRESTRIAL THERMODYNAMICS
THE EARTH A THERMAL MACHINE

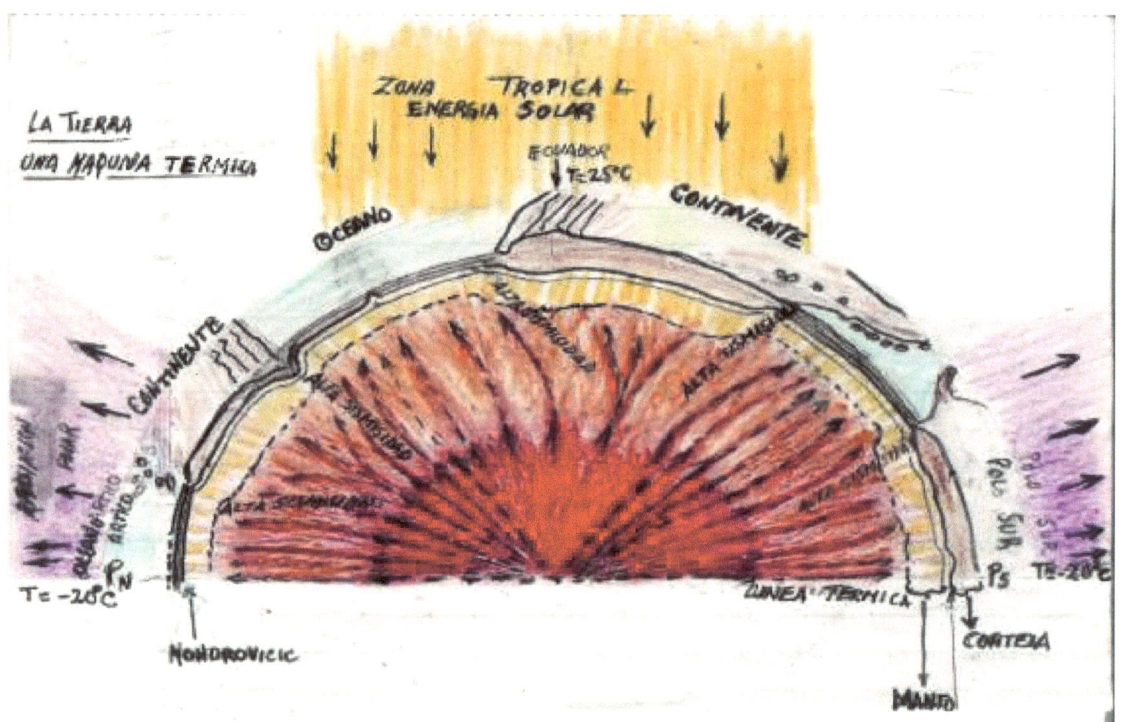

a) " Mohorovichic Discontinuity"

The discontinuity of Mohorovicic (or simply Moho) is the boundary that separate lithosphere (earth's crust) from mantle; it marks an abrupt change of seismic waves speed. When seismic waves P (Pressure waves) cross this discontinuity, its speed increases from 7.0 to 8.1 Km by sec.

The Croatian seismologist ANDRIJA MOHOROVICIC (1857-1936), working in Yugoslav seismographs in 1909, was first observe this surprising geological phenomenon.

Under the continents, Moho marks the transition between continental granite rock (SIAL) and the ultra-basic peridotite (SIMA) of the terrestrial mantle. Mohorovicic discontinuity is at an average depth of 35 km under the continents, but the depth is greater under mountain ranges.

Usually, under the ocean basins, Moho is only 6 km deep and marks the transition between the oceanic basalt of the crust and the mantle peridotite.

b) MOHO, last energy equivalence limit

Seismological studies show that MOHO is the MANTO's outer limit. The Lithosphere, rest on the MANTO with its variable form and depth.

The energy that flows from the Earth's nucleus is distributed in radial form through the sphere. It is congruent to accept that energy is distributed homogeneously through the terrestrial sphere because there is no way to verify the existence of obstacles that orient or limit its flow towards the surface; except those that crust have, as its diversity of phases, diverse energy degree, geographic position or elements with which she is in contact as atmosphere, and oceans.

Based on metallurgy concepts, we can say that a certain material changes phase when the temperature reaches the transformation thermal limit of that material. Therefore, it is legitimate to assume that the part of the mantle that contacts the lithosphere has a similar temperature over its entire surface, although the depth and geographical position vary. (See GRAPH # 1 - THE EARTH, A THERMAL MACHINE)

In a homogeneous material, similar temperature means similar energy quantum per volume's unit; consequently MOHO is the last terrestrial equal energy limit from which the energy that escapes constantly from the Earth center, crosses the different materials of the continental and oceanic lithosphere.

c) Planet Earth, a thermal machine

The Earth works like a heat machine whose energy source is in the inner core, the energy gradient between cores to crust causes the energy to flow in spherical-radial form towards the globe surface; the energy generated inside flows outside by conduction.

As we said, the discontinuity of Mohorovicic is the last zone of "equi-energy" after which the internal energy must crosses the earth's crust, to be expelled out of the planet solid part.

Therefore, it is legitimate to say that: the amount of energy radiated by Earth, from its own internal energy (without considering what received from the Sun) is:

Irradiated Q = Q" irradiated by Continents + Q' irradiated by Oceans

In addition, it is congruent to say that; from the discontinuity of Mohorovicic, the "energy radiation by area unit" would be the same in all the discontinuity surface if, and only if, the crust were homogeneous.

However, the continental masses are diverse in thickness and composition, the crust under the oceans is thinner than that of the continents and dissipates the energy it receives from sphere's interior through the ocean water whose "specific heat" it is far greater than in the continents, whatever their composition. In other words, the "energy dissipated per unit area" is not the same on the entire surface of the discontinuity because differences in refrigeration processes.

$$\frac{Q'' \text{ (radiated by continents)}}{A'' \text{ (Earth continental area)}} \neq \frac{Q' \text{ (radiated by oceans)}}{A' \text{ (Earth oceanic area)}}$$

Another modification factor is the energy received from outside (Sun). We know that the dissipation of energy is linked to "energy gradient" existence. It is obvious that solar energy that reaches the Earth tends to decrease or, in fact, to cancel the gradient between terrestrial interior and the atmosphere; the consequence is that the surplus of internal energy is stored in discontinuous and non-homogeneous form in the terrestrial interior.

As we saw, energy flows in a homogeneous way from the center of the earth through all its spherical volume until it reaches the Discontinuity of Mohorovicic, from that point it finds several obstacles that prevent it from reaching uniformly to the surface. When crossing Mohorovicic, the energy flow per unit area is diverse, as it crosses crust areas of varying thickness or combined areas of crust and oceans.

d) Earth morphological answer

The coefficient of "specific heat" is the amount of heat necessary for the unit of mass, of a given material, to increase or decrease its temperature by one degree.

The limits and characteristics of crust materials, its distribution, the oceans existence, mountain ranges, deserts and other composition cortical forms with different "specific heat" coefficient condition and modify the thermodynamic response of the Earth.

As a direct consequence, the thermodynamic response of the Earth is expressed as:

$$c = \frac{Q1 - Q2}{m \cdot \Delta T}$$

Where:
c.- Specific heat of any material
Q1.- Heat amount in a material at a given temperature
Q2.- Heat amount of the same material with 1 degree more
ΔT.- Temperature difference (1°)
m.- Amount of mass of studied material

Now, each material has its own specific heat, water (c) is the greatest of all.

From the **Table of certain common materials** data we can see that there is an important difference between the specific heats of sea water and lithosphere; and if that data is applied to the formula (Q = m c Δ T), the resultant shows significant differences in heat amount of this two materials.

Table of certain common materials of "c" (Specific heat)		
Water	1.00	cal/g°C
Seawater	0.95	cal/g°C
Copper	0.093	cal/g°C
Aluminum	0.21	cal/g°C
Oil	0.31	cal/g°C
Iron	0.115	cal/g°C
Glass	0.20	cal/g°C
Steel	0.11	cal/g°C
Platinum	0.032	cal/g°C
Asbestos	0.20	cal/g°C
Human Skin	0.36	cal/g°C
Human Body	0.86	cal/g°C
Lithosphere	0.178	cal/g°C
The real interest data to this work are:		
Seawater	0.95	cal/g°C
Lithosphere	0.178	cal/g°C

On the other hand, we know that average depth of Mohorovicic under the oceans is 'approximately' 6000 meters, and the average depth of oceans is 'approximately' 6000 meters. Consequently

The average of oceanic Q_1, per volume unit, is:

$Q_1 = (1 cm^2$ x 600,000 cm) x 1.03 gr/cm^3 x 0.95 cal/gr/°C x 1° C = 587,100 cal.
The Q_2 per volume unit, for average lithosphere from surface to Mohorovicic, is:
$Q_2 = (1 cm^2$ x 600,000 cm) x 2.7 gr/cm^3 x 0.178 cal/gr °C x 1° C = 288,360 cal.
The total Qt for this area will be:
$Q_t = Q_1 + Q_2 =$ 587,100 + 288,360 = 875,460 cal.

On the other hand, we know that the average thickness of the continental crust up to Mohorivicic is 'approximately' 36,000 meters.

Consequently, the continental Qt per area unit will be:

$Q = (1 cm^2$ x 3'600,000 cm) x 2.7 gram/cm^3 x 0.178 cal/gram/°C x 1 °C = 1'730,160 cal.

In other word, each theoretical cube of 1 cm^2 of cross section that has its base in Mohorovicic discontinuity and extends, through the lithosphere until the sea surface, accumulates an amount of heat equivalent to 875,460 calories for each degree Celsius of temperature.

On the other hand, each theoretical cube of 1 cm^2 of cross section that arises in the discontinuity of Mohorovicic and extends to the surface of the continents accumulates an amount of heat equivalent to 1'730,160 calories.

As we can see, the accumulate heat in continental areas is approximately double than that accumulates in the oceanic areas.

When comparing the capacity of energetic accumulation in continental zones with the one of oceanic zones it could be deduced that, after the energetic saturation of crust's similar volumes, the excess energy flows will be oriented towards zones of better "thermal conductivity" or with better "factor of refrigeration",

It is interesting to note that the well-known "fire belt", which surrounds the Pacific Ocean, has on its Eastern side (coast of America) a succession of mountain ranges very close to the littorals of the three continental parts that make it up (North, Center and South Am) and on its eastern ocean side, mountain ranges that run from North to South through Asia, Philippines, other archipelago, and Australia.

But we must consider also that, due to the great dimension of Asian continent and its East-West orientation, the Asian front of the "Circle of Fire" is greatly affected by the friction between Mantle and Crust. Do not forget that the Earth Globe moves from West to East and, while turning on itself, any change in the rotation speed will cause slippage between the enormous continent and the part of the mantle on which it rests. Remember the "magma-tectonic inertial slip".

Therefore, it must be assumed that the differential accumulation of energy, which flows from within, is greater in these fronts than in any other place on earth; this would be consistent with the fact that the higher incidence of earthquakes suffered by these areas is due to their geological condition and the creep and accumulation of energy that affects them.

In other words, it is also consistent accept that the differential accumulation of mass / energy began with the lithosphere formation and, after millions years, led to the division of the lithosphere into the pieces that constitute the continents today.

In the stable dynamic equilibrium time, the excess of sub-cortical mass / energy which flows to the oceans through the most favorable part of the oceanic crust, does so through oceanic ridges and other geological accidents such as underwater volcanoes; on the continents, through terrestrial volcanoes.

Remember that the mass/energy leakage weakens the attraction force (gravity) and leads to the globe growth with its sequel of lithosphere thickening and all kinds phenomena (geological, volcanic, mechanical, chemical, etc.).

In consequence we can repeat "the mass/energy loss is the engine of expansion".

2) MAGNETIC FIELD and energy leak

We know that Earth is almost spherical, rotates at a rate of one complete revolution over itself in a little more than 24 hours, and has two poles that determine the rotation axis and the existence of the Magnetic Field.

a) Earth Magnetic Field

Science states that a magnetic field is originated by the movement of electric charges. In the Earth, the massive electric charges are originated in the earth's hyper-energetic center and they moving towards the poles responding to the thermal gradient; this generates the earth's magnetic field.

At the beginning of our planet (when the lithosphere was forming, the distance to Sun was less, and the Earth's declination was almost zero) the energy flowed from the earth center in radial direction uniformly distributed towards outer space but, at reach the planet surface it found resistance. That was especially in the area with direct sunlight.

In this condition, the energy received directly from the Sun established a wide area in which the surface material retained a temperature higher than that existing at the poles.

The natural response was that the energy reoriented its flow towards the poles due to the lower resistance and greater thermal gradient. This thermodynamic condition and the mobility of the internal electric charges gave rise to the formation of the Earth's magnetic field (the process is similar in all celestial bodies that have a magnetic field; the greater or lesser magnitude of these fields obeys to: the greater or lesser mass / energy, the greater or lesser of internal electric charges mobility, the greater or lesser rotation speed)

Thanks to artificial satellites it can be said that the Earth's magnetic field has the shape of a gigantesque raindrop whose end extends far away from the Sun, towards outer space; its core is like a powerful magnet bar or a large electric coil in the planet center through which flows electricity.

Many geophysicists agree that the magnetic field is generated by electric eddies. These eddies are driven by the heat released by radioactive elements in the Earth's outer core; electro conductor and rich in iron.

When the supersonic solar wind particles hit the Earth's magnetic field, a shock wave is produced it makes evident the magnetopause that is the magnetosphere limit. The radiation belt (discovered by Van Allen) blocks the particles entry through the magneto-pause; some sneak through the open space that, during the polar summer, leaves the magnetic field lines near the poles. The solar wind particles collide with the ions' jet and atoms that leave towards the ionosphere producing the known "auroras".

b) Mass and energy leak

The greatest Earth's mass and energy leak occurs in the Magnetosphere through the Magnetopause, impacted constantly by solar wind; this leak is the biggest factor of planet energy imbalance. The mass and energy leak is constituted by a jet of ions and atoms dragged into outer space. The magnitude of the mass and energy loss is small compared to the planet dimension and does not seem to be factor that can produce an important energy imbalance.

However, the Earth has more than 5000 million years losing mass and energy; considering this and the current state of ecology, the evident climate change, the extinction of living species and others, it seems that the planet is going to a thermo-energetic unbalance due to factors other than the natural process.

3) TIDES AND ITS EFFECTS

The force of attraction exerted on the Earth by the Sun and the Moon is reflected in specific phenomena such as tides. Man has witnessed the variation in the levels of seas and oceans at different times of day or night, and had could verify that this variation was greater or lesser in a repeated cycle, coinciding with the movements of the Moon. The human being called this daily alteration of the level of seas and oceans.

With passing centuries was evident that the water level variability was connected with the moon mobility almost magically; this effect was greater in the rare days when the moon aligned with the sun in the awesome show of an eclipse.

Many boats and missing men were the price that the tribe had to pay to understand and respect the tides and variations of the sea King genius. Man gave himself a thousand explanations and all ended up assigning to Moon magical powers, his conclusion was to respect the Moon and guard against.

Established the universal attraction principle, it was possible to understand that the cyclic variation of sea level is also a universal gravitation consequence, and tides variability obeys to the position of the Moon and the Sun in front the Earth.

a) The Super Seasonal Tide

Throughout the year the earth revolves around the Sun, at a speed of 29.6 km per second, and invests 365 days, 5 hours, 48 minutes and 46 seconds in this trip. The plane of the ecliptic and that of the equator form an angle of 23.5°. This produces appropriate conditions to generate of the SUPER SEASONAL TIDE. In other words, as of December 21, the oceanic mass begins its seasonal shift towards North; on the same date on which the Sun begins its cyclical return to the North. The cyclic seasonal wave will reach the maximum north declination on June 23. The trip of the liquid mass towards the oceanic North spaces it will produce oceanic level increase in that hemisphere.

The average level of the North Pacific Ocean increases an average of fifty centimeters between the months of June to September; that same effect although in smaller magnitude happens in the Southern Hemisphere between December and March.

As is logical, the force that causes the tides affects the whole and the parts. This means that the Sun and / or Moon attraction affects the entire planet and each of its parts, oceans, seas, earth's crust and atmosphere.

Due to gases nature and existence of other factors, it is difficult to evaluate the effect produced by the lunar and solar attraction on the atmosphere, however it has been verified that the atmospheric tide affects the pressure. *(Atmospheric tides, since atmospheric air is a fluid, as are the ocean waters, the atmosphere dimensions also suffer the action of tides, affecting their thickness and height and, consequently, the atmospheric pressure.)* WIKIPEDIA)

The lunar and solar attraction on liquid elements (oceans, seas, lakes) has been widely studied, described and understood, especially in its mechanical and hydraulic effects.

The effects of lunar and / or solar attraction on solid parts, such as the crust, were ignored until very recently, but science has found that crust also suffers the solar and lunar attraction consequences.

1. The lunar and solar attraction over the solid parts have been little considered and studied. (Terrestrial tides: the gravitational forces that cause the ocean tides also deform the earth's crust. The deformation is important and the amplitude of the terrestrial tide reaches about 25 to 30 cm in live tide and almost 50 cm during the equinoxes) (WIKIPEDIA) Masselink, G.; Short, A. D. (1993). «The effect of tidal range on beach morphodynamics and morphology: a conceptual beach model». Journal of Coastal Research 9 (3): 785-800. ISSN 0749-0208.

4) "EL NIÑO" PHENOMENON

As we have already seen, the seasonal super-tide is fed by an enormous mass of water that moves from the Southern Hemisphere to the North, and increases the North Pacific ocean level; this huge water mass it is permanently heated by solar energy during its annual cyclic trip.

This mass and its important kinetic action become evident when on June 23 it starts its cyclic movement towards the south, attracted by the solar declination change. Its greater level of about 50 centimeters induces, in the ocean, a massive countercurrent that stops the South cold waters flow, until the oceanic level takes an average dimension.

In past time; the accumulated energy, that it was of enormous magnitude, moved harmonically with the oceanic mass, from northern hemisphere to South in an annual ebb and flow innocuous; because, the planet tools for natural convection process were sufficient for a balanced thermal transfer.

At that time, this thermo-mechanical-hydraulic phenomenon was neither observed nor considered because, for the primitive hydrographic science control systems, its influence on the atmosphere and climate was unimportant.

However, since pre-Hispanic times Peruvian artisanal fishermen observed, that every certain number of years in December, coastal waters was warmed above normal. When Christian-Catholic culture arrived, those fishermen said that the phenomenon arrives with the "Niño Dios" (Child God). Peruvian oceanographers of fifties in past century; impacted by the growing economic importance of fishing industry, they studied the phenomenon and baptized it with the suggestive name of "Niño Phenomenon".

It is regrettable that the Peruvian fishing company distorted the remarkable observation of artisanal fishermen by delaying and obscuring the correct interpretation by economic-private interest.

However; the increase in sea temperature, caused by the 'greenhouse effect' added to the old "El Niño" phenomenon, affects the 'thermodynamic' harmony of this super tide and increases the thermal imbalance that suffers our world today. The unbalance will increase as long as the global private interests, as in case of Peruvian fishing industry, keep diminishing its importance and hiding the effects.

5) HUMAN ACTION

The science and industry development has given human being unsuspected powers. Today humanity can alter the earth surface thermodynamic equilibrium, in the future human being power it will be greater.

The human tools used to alter the terrestrial balance are diverse but, most of them lead to the manipulation, conscious or not, of some factors of terrestrial thermodynamic equilibrium.

a) The Ozone Layer

Ozone is a gas whose molecule is made up of three oxygen atoms instead of two that make up normal oxygen. This gas is found in the atmosphere in small proportion. When in ancient past, oxygen enriched atmosphere and replacing other gases; an ozone subtle layer showed up in the upper part of the atmosphere and remains in dynamic balance till our days. Under natural conditions ozone is created and destroyed continuously but its balance remains with small fluctuations caused by solar activity or volcanic eruptions that increase gases that destroy ozone.

b)   Science recognizes the Fluorocarbons and similar gases danger

After debates and arduous studies, it was accepted that from 1960, the greatest damage to ozone layer comes from the indiscriminate use of fluorocarbons gases and other industrial chemicals.

As the study of this phenomenon has deepened, it was been concluded that the most dangerous and influential elements are: chemical products used in refrigeration, plastics industry, electro-industry's solvents, fire chemical extinguishers, gases' leaks from automotive engines, supersonic aircraft and, to which we must add the natural impact of terrestrial life.

The most impact ant chemical products for atmosphere are: bromides, nitrous and nitric oxides, carbon dioxide, chloral-fluoride-carbons, methane and others; they are all greenhouse gases that reheat the low atmosphere and increase the pressure.

c)   Ions escape by 'Polar Chimney'

The natural renewal of atmospheric gases is a balanced process between gases production, gases transformation and the gases escaping into the stratosphere in direction of outer space. The escape to the stratosphere occurs through the 'polar chimney' that is the open space left by the earth's magnetic field. The solar wind sweeps the outside of that opening (the mouth of the chimney) at a variable speed higher than 500 km / sec but can reach more than 1000 km / sec, as we seeing before.

On the other hand, greenhouse gases react chemically with oxygen and OZONE producing decrement the natural production of ions; this reduce the earth's shield power against the infrared and ultraviolet rays and produces an increase in the Earth's surface temperature. By other side, heavy gases such as carbon monoxide and dioxide and others saturate the lower atmosphere, this makes surface cooling difficult with the consequent increase in temperature. The consequence is: earth global warming.

6) GREENHOUSE EFFECT

In scientific discussion and even in common speech, referring to the Greenhouse Effect is now commonplace. The percentage increase in hazardous atmospheric components, such as carbon dioxide or carbon fluoride, has triggered the average temperature of our gaseous bubble, at least one degree Celsius higher than it was at 20th century beginning.

To this, must be added the contamination of the oceans with all kinds of garbage: petroleum and derivatives, detergents, industrial waste, plastic waste, tailings of poisonous minerals; all this, unfortunately, it leads to decrease the oceanic thermodynamic conductivity with the consequent abnormal calories accumulation in the oceanic mass.

Gigantic terrestrial areas under business ecological alterations and uncontrolled deforestation, destined for to urban planning, mining, oil exploitation, uncontrolled logging and many others; they are leaving our planet without capacity for induced thermodynamic imbalance counteracts. Those actions have as sole purpose satisfy private economic world appetites. (See Graphics # 5 PLASTIC GARBAGE IN THE OCEANS – and -graphic # 6 OCEAN POLLUTION)

a) Acid rain
In sixties and seventies past century, the most powerful countries of Northern hemisphere alerted the world against "acid rain" damage, however the growing economic-political-military power competition made it impossible to take any preventative action. As a consequence, by the eighties it was evident large damaged wooded areas in England, Germany, Czechoslovakia and Poland.

My first visit to USA was in 1956, after six decades I can affirm, without being a forestry engineer that, Californian, Oregon or Washington State's forests are not as healthy as they used to be. As a nature lover I have seen many trees with spring yellow leaves, that show is repeated in Virginia and Maryland. Trees regenerate them self if the atmosphere is improved; it is still time!

b) Deforestation.

Millions of hectares of forest are slaughtered every year in Amazonia, India, Bangladesh, Africa, Central America, and North America. This has the sole purpose of support working huge industrial complexes, operating with wood, chemical elements, chemical and pharmaceutical industrial inputs, gold, precious stones and drug. (See Graphic # 2 DEFORRESTATION)

GRAPH # 2
BALANCE IN TERRESTRIAL THERMODYNAMICS
DEFORESTATION

Ales Krivec If you like my images, please do consider buying me a cup of coffee.

> ***Nearly a third of the terrestrial farm camps have been abandons because it has rendered unproductive by erosion.***
> *Each year 20 million hectares of farmland are degraded to such an extent that they become unproductive for farming, or are lost by disorderly urban expansion.*
> *Desertification is increased to some extent by agricultural land expansion: 30 percent of irrigated land, 47 percent of which is irrigated by rain in agricultural regions, and 73 percent in hillside land.*
> *Annually, an estimated of: 1.5 to 2.5 million hectares of irrigated land, 3.5 to 4 million hectares irrigated by rain in agricultural regions, and about 35 million hectares of agricultural hillside lands lose all or part of their productivity due to the soil degradation.*
> *The restoration of soil lost by erosion is a slow process, to form a layer of 2.5 centimeters of agricultural soil, of terrestrial dimension, can it take 500 years. (excerpts from UN reports and other agencies)*

cc)

c) Irrigations

In response to the processes of desertification and population increase, the human being has tried to extend the cultivation areas through complex irrigation systems, these immense and growing areas consume a lot of groundwater, the monoculture prevents the natural recovery of soil and it requires artificial enrichment with fertilizers that are accompanied by the need for potentially dangerous pesticides. (See Graphic #3) AGRICULTURE AND DESERTIFICATION – and - # 4 DESTRUCTIVE IRRIGATIONS

GRAPH # 3
BALANCE IN TERRESTRIAL THERMODYNAMICS
AGRICULTURE AND DEFORESTATION

Photo by Marcin Kempa on - Copy

The consequence of this complex and aggressive machinery has long exceeded the natural planet recovery capacity, all of which is reflected in dreadful United Nations statistics and other international bodies of absolute credibility.

In other words, companies that exploit large agricultural areas, without clear and legally defined responsibility about the degraded soils, would spend more of the accumulated capital and work 500 years to restore, only 2.5 centimeters of the destroyed agricultural land.

These dreadful data are responses to climate change or human action or more likely to both. On a global scale, the increment in greenhouse gases increases the drought year's frequency.

Desertification maps, common in school books, show the distribution of semi-arid zones, deserts and areas in accelerated desertification process and, as can be seen, almost all those areas are among the tropics where solar energy is strongest.

GRAPHIC # 4
BALANCE IN TERRESTRIAL THERMODYNAMICS
DESTRUCTIVE IRRIGATION

Rural poverty and desertification
When land, in arid regions, is fragile and over-exploited by the demand of a growing population, that land loses its productive capacity. Today, that degraded lands affects more than 1 billion people and 40 percent of the earth's surface. In most severe cases, land becomes sterile and useless, precipitating hunger and drought. Each year 12 million hectares of land are lost due to desertification, and the rate is increasing. Desertification is a major environmental problem that is advancing at an alarming rate. (excerpts from UN reports and other agencies)

However, it is noteworthy that there are vast areas under desertification process of n Asia and North America where, the continental crust dimension makes it difficult the energy leakage, from earth interior. The desert areas distribution and its increment have a great influence on terrestrial thermodynamics.

7) GREENHOUSE PHENOMENON CONSEQUENCES

Throughout this work, we saw that the thickening of the earth's crust, the oceans existence, the atmospheric oxygen increment, other atmospheric process and the proper distance between Earth and Sun led to our planet to a state of energy balance that facilitated life; we also know that the balance was periodically altered by isolated events such as the KRAKATOA explosion and more or less destructive earthquakes.

Since the advent of the industrial age, the energy balance's dynamics have changed and this change is leaving its mark through multiple parameters, apparently disconnected.

GRAPHIC # 5

BALANCE IN TERRESTRIAL THERMODYNAMICS

PLASTIC GARBAGE IN THE OCEANS

From Wikipedia, the free encyclopedia
The plastic soup is located in the oceanic turn of the North Pacific, one of the five great oceanic turns.
The Plastic Soup. 1.- also known as Garbage Soup, Toxic Soup, Great Pacific Garbage Patch, Large Pacific Garbage Area, Pacific Garbage Swirl and other similar names, is an ocean area covered by marine debris in the center of the North Pacific Ocean, located between the coordinates 135 ° to 155 ° W and 35 ° to 42 ° N. It is estimated to have a size of 1,400,000 km². 2, - This ocean waste dump is characterized by exceptionally high concentrations of suspended plastic and other debris that has been trapped by currents of the North Pacific gyre (formed by a vortex of ocean currents). In spite of its size and density, the batch of oceanic garbage is hardly visible through satellite photographs and it is not possible to locate it with radars. In 2009 the North Atlantic Garbage Spill was discovered, also related to the North Atlantic Ocean Turn.

The world population explosion multiplied the demand and led to natural goods over-exploitation, all kind of animal species extinction, depletion or significant oil reserves reduction, irremediable contamination of river basins and inland seas. All that aggression has reached the oceans today.

Medical health improvement produced what we call world population explosion

Year	Month	Day	World population
1900	December	31	1,657'000,000
1950	December	31	2,548'600,000
2000	December	31	6,165'700,000
2015	December	31	7,391'200,000
2020	December	31	7,797'810,000
2030	December	31	8,535'600,000

a)   Thermodynamic Balance between Crust and Atmosphere

Twentieth century last years passed quickly while, in international conferences, newspapers, magazines and family conversations were openly commented on atmospheric pollution, disappearance of atmospheric ozone and the irrepressible skin cancer increase, as a result of ultraviolet rays whose passes through our weakened atmospheric shield.

On the contrary, almost nobody thinks about the fact that; any terrestrial surface temperature variation affects also the internal thermodynamic equilibrium.

It is a fact that energy flows from the earth's center to the surface of the planet; in that condition, the internal energy is opposed to the one coming from the sun and so a thermodynamic equilibrium is established, this govern thickness of the earth's crust and, of course, the depth of " de Mohorovicic Discontinuity".

It is possible to verify also that each differential degree in the surface temperature implies an enormous amount of calories that the bark accumulates, according to its own nature, modifying, controlling and affecting the internal energy flow.

If it were possible isolate these phenomena, it would be verified that: the increase of one centigrade degree in Earth surface reduces thickness of the crust between 11 and 100 meters; this of course, with its sequel of potential increase in tectonic activity. That extra centigrade degree on the surface of the earth will increase the transfer of energy from the center to the poles and will strengthen the movement of electric charges because it alters the magnetic field and the lines of force; therefore the "siphon effect" will increase over our atmosphere with the consequent increment in loss of light atmospheric gases such as ozone, oxygen and others.

An additional Celsius degree, on the globe surface, will accumulate more thermal energy in the oceans, which will affect the global aquatic mass in its transit from north to south and vice versa; this, of course, will alter the boundary between cold and warm water with the consequent dry-wet and hot-cold cycles modification.

One average additional centigrade degree, in world surface, alters the thermal gradient between Poles and Equator and will affect the polar ice whose decrease will be progressively faster as lower its volume; this is happening today in Andean peaks, South Pole and many other places in our world.

The average surface temperature's increment breaks the thermodynamic equilibrium of the terrestrial globe and, with time, this trend will produce an uncontrollable increasing process since all the conditions reinforce that tendency and there are not reduction or control factors.

b)   Hazards generated if the Thermodynamic Balance is broken

It is demonstrated that, from 1900 to now, the average of Earth surface temperature has increased approximately one Celsius degree and, for 2020 the same studies calculate three possible projections (See Graphics # 5 y # 6):

Optimistic projection. - The emission of greenhouse gases, deforestation, development of commercial aviation and many other human aggressions against our world will be controlled by responsible governs. Under these conditions, the average global temperature will increase half Celsius degree and, if this control trend is maintained, it is expected that the year 2100 the thermodynamic terrestrial condition will be similar that our present time. By that date the bark thickness will be one hundred (100) meters less thick than in 1900.

Realistic projection. - The human beings have, today, awareness about climate change; this allows us to understand the phenomenon and, in the capitalist competition free play, humanity could achieve a restricted increment of aggressive factors. Under this condition; by 2020, the global temperature average could increased one degree Celsius and, if that trend is maintained throughout the century, we could expect in year 2100 for our great-grandchildren, two Celsius degrees above the current temperature. By that date the thickness of the bark will be 300 meters less than in 1900.

Pessimistic projection. - The pessimistic projection foresees that, all aggressive phenomena reinforce each other; the competition between production centers becomes uncontrollable, the human mass attacks the remnant natural redoubts (Amazonian basin, African and Asian rain forests) and as a result of the irrepressible pollution, oceans, jungles and forests decreases its oxygen production.

Under these conditions; by 2020 the global temperature will have increased one and a half Celsius degree and, it will surely reach five centigrade degrees above the temperature of our days, towards 2100. For that date the crust will be, on average, six hundred (600) meters less thick than in 1900.

c) Consequences of Thermodynamic Balance Breakdown

We saw in past pages that, the impact to increase one Celsius degree is pretty dangerous. In that sense if the surface average planet temperature reaches five degrees higher than at present, this increment will project uncontrollable dangers, will irremediably affect our world and, of course, human life:

GRAPHIC # 6
BALANCE IN TERRESTRIAL THERMODYNAMICS
OCEAN POLLUTION

- THE DEPTH OF MOHO, Mohorovicic discontinuity will approach to planet surface because the crust thickness will decrease 600 meters with its sequel of tectonic instability that would reach frightening levels;
- THE ENERGY TRANSFER TOWARDS THE POLES, ionized gases leakage into outer space will increase and affect not only the ozone layer but will endanger the proper atmospheric oxygen balance;
- ENERGETIC GRADIENT BETWEEN THE POLES AND ECUADOR, the change in the energy gradient will profoundly alter Earth's climate with special incidence in the well-known PHENOMENA OF THE "Niño" and "Niña" and with them the incidence of hurricanes, droughts, monsoons and alteration of oceanic currents;
- INTER FACTOR REINFORCEMENT, with the increase crust temperature due to changes in the atmospheric gases, the sphere mass (nuclei, mantle and bark) will accumulate thermal energy, will increase the thermal gradient between the nucleus and the poles, ices will melt in all around the world, it will affect earth gravity and will make unpredictable the response of the planet which has given everything to its inhabitants and only asked us respect the laws that govern its dynamic-energetic balance;
- MOHO ALTERATIONS WILL GENERATE IMBALANCES, if the temperature increase lasts time enough to melt 1000 crust meters and so reduce the distance between MOHO and Earth surface, the consequences will be extremely dangerous (we must be remembered that MOHO is, on average, only 6,000 meters below the oceans).

By increasing the thermal difference between continental crust and oceanic crust, due to their diverse specific heat, the tectonic mobility in the contact zones will increase and earthquakes incidence and magnitude will substantially increase.

On the other hand, the influence of solar attraction on tectonic activity increment in seismic zones is proven. Consequently, it is logical to expect greater crust deformations and increase of tectonic activity in response to the greater Moho proximity to planet surface, and as a consequence of the annual Sun's go and back (Sun goes from 23 North to 23 South and vice versa during the year).

All said has a cause - effect relationship with the Earth surface temperature increase, produced by greenhouse phenomenon.

EPILOGUE

It is proven that the Greenhouse Phenomenon, ozone gap, desertification, soil degradation, oceans and atmosphere pollution, added to other degrading factors increase the globe surface temperature and, as this work aims to demonstrate, the earth's crust is also seriously affected. Very few members of humanity enjoy this aggression against our only one universal home while humanity as a whole suffers it and will suffer more and more in a very near future.

www.ingramcontent.com/pod-product-compliance
Lightning Source LLC
Chambersburg PA
CBHW051913210526
45473CB00006B/1991